Math Mammoth Division 2

By Maria Miller

ISBN 978-1-954358-72-0

Contents

Introduction

Math Mammoth Division 2 is a continuation of the book *Math Mammoth Division 1*. It focuses on the topics of long division, remainder, problem solving, average, divisibility, and factors. The book is most suitable for fourth grade.

We start out by reviewing basic division facts by single-digit numbers (such as 24 ÷ 4 or 56 ÷ 7). After that, we study terminology of division and dividing numbers by whole tens and hundreds (such as 400 ÷ 20). Next students practice the order of operations—this time with division as one of the operations.

Then we study the concept of remainder, preparing students for the upcoming lessons on long division. At first, the concept of remainder is presented visually. Soon, students solve simple division problems with a remainder, written with the long division symbol (or long division "corner", as I like to call it).

Next comes a set of lessons intended to teach long division in several small steps. We start with divisions where each of the digits in the dividend (thousands, hundreds, tens, and ones) can be divided evenly by the divisor (for example, 3096 ÷ 3). As the next step, there is a remainder in the ones. Then, the divisions have a remainder in the tens. Finally, there is a remainder in the hundreds and in the thousands, and this completes the step-by-step learning process for long division. The lessons also include lots of word problems to solve.

After long division, we study the concept of average, which is a nice application of division, and problems that involve finding a fractional part of a quantity using division. For example, we can find 3/4 of a number by first finding 1/4 (dividing by 4) and then multiplying the result by 3. Students get help from visual bar models to solve the problems.

The last section deals with elementary number theory. We study basic divisibility rules (though not all of them), prime numbers, and finding all factors of a given two-digit number.

Answers are at the end of the book.

I wish you success in teaching math!

Maria Miller, the author

Helpful Resources on the Internet

We have compiled a list of external Internet resources that match the topics in this book. This list of links includes web pages that offer:

- **online practice** for concepts;
- online **games**, or occasionally, printable games;
- **animations** and interactive **illustrations** of math concepts;
- **articles** that teach a math concept.

We heartily recommend you take a look at the list. Many of our customers love using these resources to supplement the bookwork. You can use the resources as you see fit for extra practice, to illustrate a concept better, and even just for some fun. Enjoy!

https://l.mathmammoth.com/blue/division2

Review of Division

Multiplication has to do with equal-size groups: 2 × 4 means 2 groups of 4.
Division is the opposite operation of multiplication, and it *also* has to do with equal-size groups:
8 ÷ 4 can mean, "How many groups of 4 are in 8?"
It can also mean, "How many in each group, when 8 things are put into 4 groups?"

Division has two "meanings":

- Dividing to find how many are in each group.
- Dividing into groups of a certain size.

2 × 6 = 12

"12 divided into 2 groups; how many in each group?"

12 ÷ 2 = 6

OR

"How many sixes are in 12?"

12 ÷ 6 = 2

6 × 2 = 12

"12 divided into 6 groups; how many in each group?"

12 ÷ 6 = 2

OR

"How many twos are in 12?"

12 ÷ 2 = 6

1. Write a multiplication sentence and two division sentences.

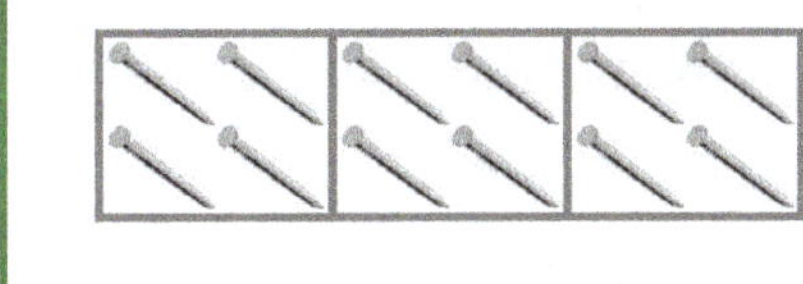

a. ____________________

____

____

b. ____________________

____

____

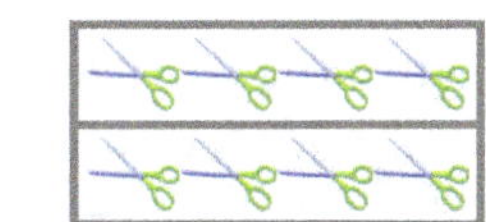

c. ____________________

____

____

2. Fact families: write two division and two multiplication sentences.

a. 21 7 and 3	**b. 24** 4 and ____	**c. 36** 4 and ____

3. Practice a little. Continue the patterns for three more steps.

a.	b.	c.	d.
16 ÷ 2 = _______	45 ÷ 5 = _______	90 ÷ 10 = _______	56 ÷ 7 = _______
18 ÷ 2 = _______	40 ÷ 5 = _______	100 ÷ 10 = _______	49 ÷ 7 = _______
20 ÷ 2 = _______	35 ÷ 5 = _______	110 ÷ 10 = _______	42 ÷ 7 = _______

4. Fill in the tables.

Eggs	6	12	24	36		54		78
Omelets	1				7		11	

Thumbtacks	8	24	32	48				
Pictures	1				8	10	12	13

5. Write a number sentence for each situation (It is not always division!). Explain what you find out from your calculation.

a. Three children shared equally 18 marbles. 18 ÷ 3 = 6. Each child got 6 marbles.	**b.** Jim has $34 and he wants a $45 book.
c. A fruit store received a shipment of 400 apples in four boxes.	**d.** Mrs. Davis divided 24 pieces of chocolate equally between 6 persons.
e. Five boxes arrived at the bookstore, each containing 50 books.	**f.** Mom bought two books that cost $13 each.
g. A herd of cows had a total of 20 legs.	**h.** Sixty books were placed on three shelves.

6. Divide.

a. $36 \div 4 =$ _____	**b.** $54 \div 9 =$ _____	**c.** $32 \div 8 =$ _____	**d.** $24 \div 3 =$ _____
$50 \div 5 =$ _____	$42 \div 7 =$ _____	$64 \div 8 =$ _____	$27 \div 9 =$ _____
$60 \div 12 =$ _____	$48 \div 6 =$ _____	$72 \div 9 =$ _____	$35 \div 7 =$ _____

7. Find what number x stands for.

a. $64 \div x = 8$	**b.** $35 \div x = 7$	**c.** $x \div 5 = 9$	**d.** $x \div 9 = 6$
x = 8	$x =$ _______	$x =$ _______	$x =$ _______

8. For each division, write a multiplication. Then find the value of the unknown.

a. $N \div 3 = 10$ ______________________________ N = _______	**b.** $x \div 4 = 9$ ______________________________ $x =$ _______
c. $60 \div T = 20$ ______________________________ T = _______	**d.** $81 \div y = 9$ ______________________________ $y =$ _______

9. Write a number sentence for each situation (It is not always division!). Explain what your answer tells you.

a. How many books can you buy for \$3 each with \$21?	**b.** A hundred apples were boxed into 5 boxes.
c. Five boxes of nails cost \$30.	**d.** A chocolate bar has 8 rows and 5 columns of squares.
e. How many fives are in 45?	**f.** Each of the five boxes weighs 12 pounds.

Division Terms and Division with Zero

Study the terms in the picture.

Notice: both the expression 56 ÷ 7 and its answer are called "the quotient"!

You can call "56 ÷ 7" the quotient written, and 8 the quotient solved.

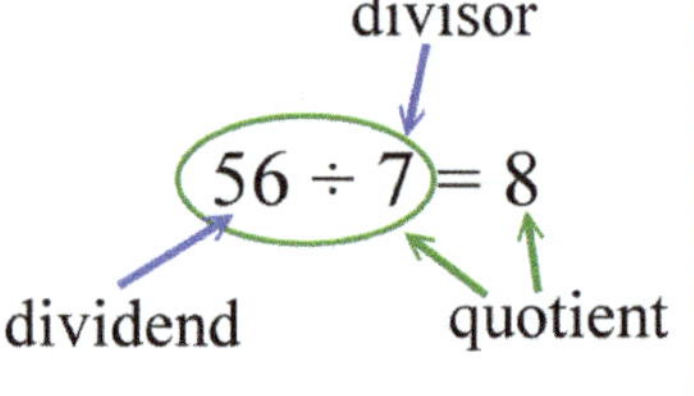

1. What is missing from these divisions: the dividend, the divisor, or the quotient? Complete.

a. 80 ÷ ______ = 40   The ______________________ is missing.

b. ______ ÷ 7 = 5   The ______________________ is missing.

c. 120 ÷ 10 = ______   The ______________________ is missing.

2. Write a division problem. Solve for the unknown.

a. The divisor is 7, the dividend is x, and the quotient is 3.   _______ ÷ _____ = _____ ; x = _____

b. The dividend is 140, the divisor is y, and the quotient is 7.   _______ ÷ _____ = _____ ; y = _____

c. The quotient is z, the divisor is 5, and the dividend is 150.   _______ ÷ _____ = _____ ; z = _____

3. Make up:

a. three division problems with a quotient of 6	**b.** three division problems with a dividend of 24
________ ÷ _______ = ______	________ ÷ _______ = ______
________ ÷ _______ = ______	________ ÷ _______ = ______
________ ÷ _______ = ______	________ ÷ _______ = ______

4. Fill in the tables.

Numbers	Product (written)	Product (solved)	Quotient (written)	Quotient (solved)
12 and 3	*12 × 3*	*36*		
10 and 5				
20 and 4				
100 and 10				

Division with zero

We check a division problem by multiplication.
Is $0 \div 3 = 0$? Check if $0 \times 3 = 0$. Yes, it is.
Is $0 \div 11 = 0$? Check if $0 \times 11 = 0$. Yes, it is.

Is $3 \div 0 = 0$? Check if $0 \times 0 = 3$. It is **not**.

Is $3 \div 0$ perhaps 3? Check if $0 \times 3 = 3$. It is **not**.

In fact, dividing by zero is a real problem. No matter what number you suggest as an answer to the problem $3 \div 0$, the multiplication check won't work, because you end up multiplying by zero, and can never get the dividend as an answer.

That is why division *by zero* is *undefined*—we cannot define a sensible answer.
You can, however, divide zero by any number (except zero). The answer is always zero.

What about $0 \div 0$?

We cannot really determine any single answer, because all of these could work:

If $0 \div 0 = 1$, then check: $0 \times 1 = 0$ works.
If $0 \div 0 = 7$, then check: $0 \times 7 = 0$ works.
If $0 \div 0 = 0$, then check: $0 \times 0 = 0$ works.

So $0 \div 0$ is usually said to be an *indeterminate* form since we cannot determine an answer to it.

Division by zero is undefined—you cannot do it.

5. Divide. Mark off the problem if it is impossible to do.

a.	b.	c.	d.
$64 \div 8 =$ ______	$55 \div 5 =$ ______	$50 \div 1 =$ ______	$0 \div 1 =$ ______
$0 \div 8 =$ ______	$6 \div 0 =$ ______	$0 \div 10 =$ ______	$1 \div 1 =$ ______
$32 \div 32 =$ ______	$7 \div 7 =$ ______	$0 \div 0 =$ ______	$9 \div 0 =$ ______

6. Find what the unknown stands for.

a.	b.	c.	d.
$64 \div x = 1$	$35 \div T = 35$	$0 \div x = 0$	$y \div 18 = 1$
$x =$ ________	$T =$ ________	$x =$ ________	$y =$ ________

7. Make up:

a. two divisions with a quotient of 1	**b.** two divisions with a dividend of 0
_______ ÷ _____ = _____	_______ ÷ _____ = _____
_______ ÷ _____ = _____	_______ ÷ _____ = _____

Mark had two division problems with the same dividend and the same quotient, yet the divisors were different. How could that be?

Dividing with Whole Tens and Hundreds

<table>
<tr><td colspan="2">Solving division problems always involves thinking
the opposite of multiplication, or "how many times."</td></tr>
<tr><td>4,800 ÷ 60 = ?

Think "backwards" of multiplication:

Because 60 × 80 = 4,800,
then 4,800 ÷ 60 = 80.</td><td>4,000 ÷ 400 = ?
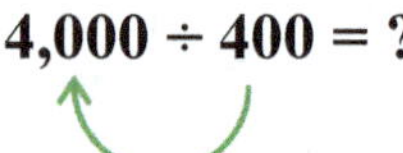
Or ask, "How many times does 400 go into 4,000?"

Ten times. So, 4,000 ÷ 400 = 10.</td></tr>
</table>

1. Solve the multiplication and then write *two* matching divisions.

a. 300 × 7 = ________	**b.** 50 × 800 = ________	**c.** 60 × 40 = ________
________ ÷ 7 = ________	________ ÷ ______ = ______	________ ÷ ______ = ______
________ ÷ 300 = ______	________ ÷ ______ = ______	________ ÷ ______ = ______

2. Think of multiplication in order to divide. Also, compare the problems.

a. 400 ÷ 8 = ________	**b.** 5,000 ÷ 5 = ________	**c.** 4,200 ÷ 700 = ________
400 ÷ 80 = ________	5,000 ÷ 50 = ________	4,200 ÷ 70 = ________
4,000 ÷ 800 = ________	5,000 ÷ 500 = ________	420 ÷ 70 = ________
4,000 ÷ 80 = ________	500 ÷ 50 = ________	420 ÷ 7 = ________

3. Solve the unknown factor problems and the division problems. Notice something special here!

a. ______ × 6 = 540	**b.** 3 × ______ = 2,700	**c.** ______ × 40 = 2,800
540 ÷ 6 = ______	2,700 ÷ 3 = ______	2,800 ÷ 40 = ______

4. Practice some more.

a. 320 ÷ 8 = ________	**b.** 540 ÷ 60 = ________	**c.** 360 ÷ 6 = ________
320 ÷ 80 = ________	540 ÷ 6 = ________	3,600 ÷ 60 = ________
3,200 ÷ 8 = ________	5,400 ÷ 60 = ________	36,000 ÷ 6 = ________

You can divide larger numbers digit-by-digit—if the division is even in each digit.
Example 1. To find 640 ÷ 2, you can divide 6, 4, and 0 by 2 (separately). Each of those is an even division. We get 640 ÷ 2 = 320. **Example 2.** We can divide digit-by-digit to solve 309 ÷ 3, because each of the digits (3, 0, and 9) can be divided by 3 evenly. We get 103.

5. Solve.

a. 426 ÷ 2 = _________ **b.** 8,044 ÷ 2 = _________ **c.** 9,303 ÷ 3 = _________

d. 550 ÷ 5 = _________ **e.** 4,008 ÷ 4 = _________ **f.** 2,820 ÷ 2 = _________

Finding half...	...is the same as dividing by 2!
$\frac{1}{2}$ of 280 is _______	280 ÷ 2 = _______

6. Find half of these numbers.

a.	b.	c.	d.
$\frac{1}{2}$ of 80 = _____	$\frac{1}{2}$ of 24,000 = ________	$\frac{1}{2}$ of 660 = _____	$\frac{1}{2}$ of 4,200 = ________

Write the numbers given in the problems into the bar models.

7. Dad used half of his paycheck to pay the rent, electricity, and water. After that he had $806 left. How much was his paycheck?

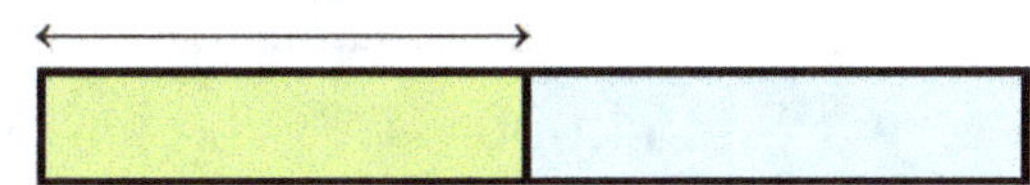

8. A fisherman sold half of his 800-kg catch to one store, and then another 350 kg to another store. How many kilograms of fish does he have left now?

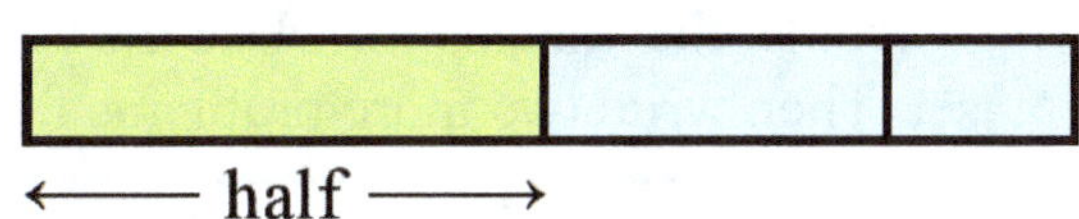

9. Emma spent half of her money to buy a phone. Then she spent $12 on some food. Now she has $15 left. How much did she have in the beginning? (You can draw a bar model to help you.)

10. Estimate. Round the dividend (the first number) so that you can easily divide mentally.

a. $352 \div 5$	b. $198 \div 4$	c. $403 \div 8$
≈ ______ ÷ 5 = ______	≈ ______ ÷ 4 = ______	≈ ______ ÷ 8 = ______

11. Estimate. This time round *both* the dividend and the divisor to the nearest ten.

a. $802 \div 21$	b. $356 \div 61$	c. $596 \div 32$
≈ ______ ÷ ______ = ____	≈ ______ ÷ ______ = ____	≈ ______ ÷ ______ = ____

12. Estimate each division result by rounding either the dividend or the divisor.

a. $80 \div 21 \approx$	b. $46 \div 5 \approx$
$120 \div 59 \approx$	$16{,}235 \div 400 \approx$
$2{,}000 \div 512 \approx$	$297 \div 30 \approx$

13. Write a division problem with a divisor of 5 and a quotient of 90.

14. Write a division problem that is impossible to solve.

15. Find what number the unknown stands for.

a. $y \times 8 = 64{,}000$	b. $s \div 6 = 700$	c. $2{,}400 \div w = 80$
$y =$ __________	$s =$ __________	$w =$ __________

16. Solve. Notice the patterns. Use the top two problems to help you solve the *third* problem in each set. Then write two more problems for each set, continuing the pattern.

a.	b.	c.
500 ÷ 5 = ________	466 ÷ 2 = ________	366 ÷ 3 = ________
505 ÷ 5 = ________	468 ÷ 2 = ________	369 ÷ 3 = ________
510 ÷ 5 = ________	470 ÷ 2 = ________	372 ÷ 3 = ________
______ ÷ 5 = ________	______ ÷ 2 = ________	______ ÷ 3 = ________
______ ÷ 5 = ________	______ ÷ 2 = ________	______ ÷ 3 = ________

Order of Operations and Division

1. Do operations within ( ) first.

2. Then multiply and divide, from left to right.

3. Then add and subtract, from left to right.

1. Solve. When there are many multiplications and divisions, do them from left to right.

$24 \div 3 \times 2 \div 4$ $= 8 \times 2 \div 4$ $= 16 \div 4 = 4$	**a.** $18 \div 2 \div 3$	**b.** $160 \div 4 \times 20 \div 8$
	c. $60 \times 20 \div 10$	**d.** $5 \times 80 \div 4 \times 20$

2. Solve. Do multiplications and divisions first.

$36 \div 3 + 20 \div 4$ $= 12 + 5$ $= 17$	**a.** $12 \times 5 + 6 \div 3$	**b.** $16 \times 2 + 15 \times 8$
	c. $80 \times 30 - 4{,}000 \div 10$	**d.** $400 \div 50 + 400 \div 40$

3. Solve what is within the parentheses first.

$(36 + 4) \div (5 + 5)$ $= 40 \div 10$ $= 4$	**a.** $(100 - 1) \div (5 + 6)$	**b.** $(140 + 17 + 13) \div 10$
	c. $2 \times (700 \div 7)$	**d.** $150 \div (13 + 7 + 10)$

4. Solve. Compare.

a.	**b.**	**c.**
$24 \div 2 + 10 =$ ________	$18 + 30 \div 2 =$ ________	$40 - 40 \div 8 =$ ________
$24 \div (2 + 10) =$ ________	$(18 + 30) \div 2 =$ ________	$(40 - 40) \div 8 =$ ________

5. Write a single number sentence and find the final answer.

a. First find the sum of 20 and 15. Then divide that by 5.
b. First find the quotient of 50 and 5. Then subtract that result from 20.
c. Find the product of 20 and 30. Then subtract 100 from it.

6. Which calculation solves the problem below?

Sharon had 21 toy figures and Joe had 17. They put their toy figures together and shared them equally. How many figures did each child get?

$21 + 17 \div 2$	$2 \times (21 + 17)$
$2 \div (21 + 17)$	$(21 + 17) \div 2$

7. Which calculation solves the problem below?

Tickets to the zoo cost $6 each. Sandra and three of her friends shared equally, the cost of six tickets. How much was each person's share of the cost?

$3 \times 6 \times 6$	$6 \div (4 + 6)$
$6 \times 6 \div 4$	$6 \times 6 \div 3$
$3 \times 6 \div 4$	$(6 + 6) \div 4$

8. Here are some interesting-looking calculations. Solve.

a. $5 \times 10 \div 10 =$ _______ $7 \times 9 \div 9 =$ _______	**b.** $60 \div 2 \times 2 =$ _______ $120 \div 40 \times 40 =$ _______	**c.** $20 \div 20 \times 20 =$ _______ $20 \times 20 \div 20 =$ _______

d. $(10 \times 10) \div (10 \times 10) =$ _______ $(10 \div 10) \times (10 \div 10) =$ _______	**e.** $(10 - 10) \div (10 + 10) =$ _______ $(10 + 10) \times (10 - 10) =$ _______

9. Use the four operations, number 5, and the parentheses to make the number sentences true.

a. 5 ☐ 5 ☐ 5 = 5 **b.** 5 ☐ 5 ☐ 5 = 0 **c.** 5 ☐ 5 ☐ 5 = 2

d. 5 ☐ 5 ☐ 5 ☐ 5 = 100 **e.** 5 ☐ 5 ☐ 5 ☐ 5 = 25

Can you make 11 number sentences whose answers are the whole numbers from 0 to 10, just by using the number 5, the four operation symbols, and the parentheses? Try it!

Puzzle Corner

The Remainder, Part 1

Sometimes we can't divide objects into groups evenly and some of the objects are left over. Those "leftovers" are the **remainder**. We mark the remainder in division with the letter R.

15 ÷ 6 = 2 R3

These 15 berries are divided into groups of 6, as evenly as possible.

We can write the division **15 ÷ 6 = 2 R3**. The divisor (6) tells us how many berries there are in each group. The answer (2) tells us how many groups we got. The remainder is 3 berries.

1. Divide the things into groups of a certain size. Write a division. There will be a remainder.

a. Divide into groups of 4.

_____ ÷ _____ = _____ R__

b. Divide into groups of 2.

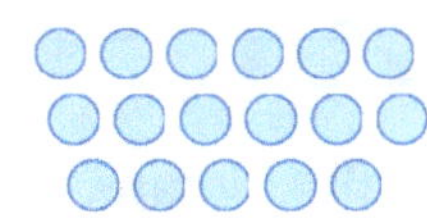

_____ ÷ _____ = _____ R__

c. Divide into groups of 5.

_____ ÷ _____ = _____ R__

2. Write a division with a remainder to match the picture. The size of the groups gives you the divisor.

a. _____ ÷ _____ = _____ R__

b. _____ ÷ _____ = _____ R__

c. _____ ÷ _____ = _____ R__

3. Draw a picture to match the division problem, and solve. Think of making groups of a certain size.

a. Divide 16 into groups of 5.

_____ ÷ _____ = _____ R__

b. Divide 17 into groups of 3.

_____ ÷ _____ = _____ R__

c. Divide 15 into groups of 4.

_____ ÷ _____ = _____ R__

4. Draw a picture to match the division problem, and solve.

a. 17 ÷ 4 = _____ R_____

b. 9 ÷ 2 = _____ R_____

c. 11 ÷ 6 = _____ R_____

Besides dividing objects into groups of a certain size, we can also divide them into so many groups.

Here you see 14 berries divided into 3 groups as evenly as possible.

We can write the division **14 ÷ 3 = 4 R2**. This time, the divisor (3) tells us how many groups we made. The answer (4) tells us how many berries are in each group. The remainder is 2 berries.

5. Divide the objects into as many groups as indicated. Write a division. There will be a remainder.

a. Divide into three groups.	**b.** Divide into five groups.	**c.** Divide into four groups.
	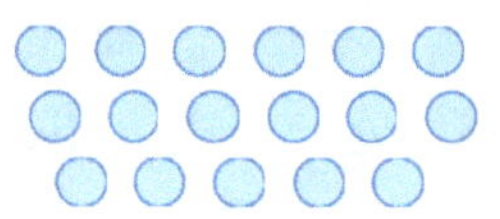	
_____ ÷ _____ = _____ R__	_____ ÷ _____ = _____ R__	_____ ÷ _____ = _____ R__

6. Write a division with a remainder to match the picture. The number of groups gives you the divisor.

a.	**b.**	**c.**
_____ ÷ _____ = _____ R__	_____ ÷ _____ = _____ R__	_____ ÷ _____ = _____ R__

Find the remainder by thinking of the DIFFERENCE.

Example. What is 35 ÷ 6?

Think how many groups of 6 there are in 35, or how many times 6 goes into 35.
You can find out with multiplication: 5 × 6 = 30; 6 × 6 = 36. So, 6 goes into 35 five times.

Now find the difference between (5 × 6) and 35, or in other words between 30 and 35.
That difference is 5, and it is the remainder. So 35 ÷ 6 = 5 R5.

7. Solve.

a. 27 ÷ 5 = _______ R___ How many times does 5 go into 27?	**b.** 16 ÷ 6 = _______ R___ How many times does 6 go into 16?	**c.** 11 ÷ 2 = _______ R___ How many times does 2 go into 11?
d. 37 ÷ 5 = _______ R___	**e.** 26 ÷ 3 = _______ R___	**f.** 56 ÷ 9 = _______ R___
g. 43 ÷ 5 = _______ R___	**h.** 34 ÷ 6 = _______ R___	**i.** 40 ÷ 7 = _______ R___

8. Solve.

a.	b.	c.
23 ÷ 4 = ______ R____	16 ÷ 7 = ______ R____	21 ÷ 8 = ______ R____
23 ÷ 5 = ______ R____	20 ÷ 3 = ______ R____	12 ÷ 9 = ______ R____

9. Divide and find the remainder. Notice the patterns!

a.	b.	c.
10 ÷ 5 = __2__ R_0_	17 ÷ 3 = ______ R___	12 ÷ 4 = ______ R___
11 ÷ 5 = ______ R___	18 ÷ 3 = ______ R___	13 ÷ 4 = ______ R___
12 ÷ 5 = ______ R___	19 ÷ 3 = ______ R___	14 ÷ 4 = ______ R___
13 ÷ 5 = ______ R___	20 ÷ 3 = ______ R___	15 ÷ 4 = ______ R___
14 ÷ 5 = ______ R___	21 ÷ 3 = ______ R___	16 ÷ 4 = ______ R___
15 ÷ 5 = ______ R___	22 ÷ 3 = ______ R___	17 ÷ 4 = ______ R___

10. Write a number sentence for each word problem. Indicate the remainder, if any.

a. Jim arranged 27 toy cars into rows of 5. How many rows did he have? Were any left over? ______ ☐ ______ = __________	**b.** The teacher put 19 children into groups of 5. How many groups of 5 did she have? What can she do with the "remainder" children? ______ ☐ ______ = __________
c. Mom baked three dozen cookies. She ate three of them, and put the rest into bags, 6 cookies in each bag. How many full bags are there? ______________________________	**d.** Jerry packaged 51 magazines into 8 bags. Was he able to do so evenly (the same number of magazines in each bag)? ______________________________
e. Susan wants to organize 35 chairs into nice even rows. Can she organize them into rows of four chairs? Rows of five? Rows of six? Rows of seven?	**f.** Amy put 38 photographs into a photo album. On each page she could fit six photos. How many photos were on the last page? How many pages were full?

The Remainder, Part 2

Division can also be written this way. The answer goes on top of the line.	This is 45 ÷ 9 = 5. $\begin{array}{r} 5 \\ 9\overline{)45} \end{array}$	This is 21 ÷ 3. Write the answer in the right place. $3\overline{)21}$

1. Divide.

a. $8\overline{)24}$ **b.** $5\overline{)45}$ **c.** $7\overline{)49}$ **d.** $8\overline{)72}$

You can also find the remainder **by subtracting**. Remember, it is the **difference**—the '**leftovers**'.	$\begin{array}{r} 7 \\ 5\overline{)36} \end{array}$ How many times does 5 go into 36? Write the answer on top of the line.	$\begin{array}{r} 7 \\ 5\overline{)36} \\ -35 \\ \hline 1 \end{array}$ Now multiply 7 × 5 = 35. Write 35 under 36. Subtract. You get 1. It is the remainder.

2. Divide and find the remainder by subtracting!

a. $\begin{array}{r} 6 \\ 5\overline{)32} \\ -30 \\ \hline \end{array}$

b. $\begin{array}{r} 5\overline{)44} \\ -40 \\ \hline \end{array}$

c. $6\overline{)37}$

d. $7\overline{)29}$

e. $8\overline{)46}$

f. $9\overline{)52}$

g. $4\overline{)35}$

h. $9\overline{)57}$

> To check your answer to a division problem with a remainder, multiply your answer by the divisor, then add the remainder. You should get the number you were dividing.
>
> **Example.** Is the division 67 ÷ 8 = 8 R5 correct? Check: 8 × 8 + 5 = 69. No, it is not.

3. Check your answers to the divisions in problem #2.

Jane packaged 27 cookies into small containers. Four cookies fit into one container. How many containers did she need?

You can divide $27 \div 4 = 6$ R3. Is the answer 6 containers, and 3 cookies did not fit?

If she puts the "leftover" 3 cookies into a container too, she will actually need 7 containers! But if she decides to eat or give away the 3 "remainder" cookies, then she only needs 6 containers.

In each question below, write a number sentence or several to show your work.

4. Jill put 33 cookies into containers. Six cookies fit into each container.

   How many containers did she need?

   How many full containers did she get?

5. Mom placed 53 eggs into cartons of 12. How many cartons did she need to hold all the eggs?

6. The teacher had 36 pencils. She divided them evenly among 11 students and put the rest of the pencils back in the cabinet. How many pencils were put in the cabinet?

7. Tom and Rich are brothers, and they have a large toy car collection—58 cars! The boys put all of the cars into boxes, eight cars per box. How many boxes were full?

8. Twenty-three children participated in a race. Their team leader gave each of them three stickers, and after that, he had 15 stickers left. How many stickers did the team leader have originally?

The Remainder, Part 3

Example 1. There were 238 children who participated in a special science class. They were divided into groups of 50. How many groups were there?

Instead of trying to divide 238 ÷ 50, you can *add* to find the answer.

50 children make one group.
100 children make 2 groups.
150 children make 3 groups.
200 children make 4 groups.
250 children make 5 groups.

This means they had four groups of 50 (a total of 200), and one other group of 38.
In other words, the 238 children were grouped like this: 50 + 50 + 50 + 50 + 38.

From this, we can write the division 238 ÷ 50 = 4 R38. Addition/multiplication will work just as well. In fact, multiplication is always used to solve division problems.

1. A hundred school children traveled to a pool in buses. Each bus could hold 42 children.
   How many buses were needed?

2. Jessica printed 73 pages of worksheets and put them into folders. Each folder could hold 20 pages.

   How many folders did she need?

   How many folders were full?

3. A school has 77 first graders.

   a. How many classes of 22 first graders could they make?

   b. The school decided to put 20 first graders in each class.
   How many first-grade classes with 20 students will they have?

   How many first graders are left over to form a class with less than 20?

4. The gym leader divided 20 players into three teams, as evenly as possible.
   How many players were on each team?

5. Divide and find the remainder by subtracting. Then, check your answers.

a.	**b.**	**c.**	**d.**
$6\overline{)47}$	$7\overline{)58}$	$5\overline{)44}$	$9\overline{)39}$

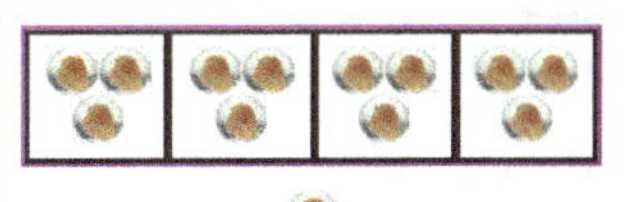	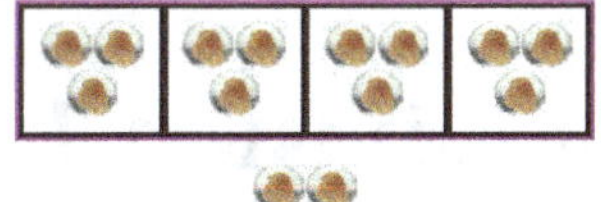	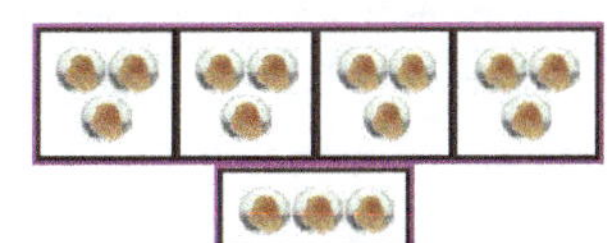
13 ÷ 3 = 4 R1	14 ÷ 3 = 4 R2	15 ÷ 3 = 5 R0
13 divided into groups of 3 makes 4 groups. One is left over.	Add one more marble. It is part of the leftovers.	Add one more. Now, instead of three "leftover" marbles, we can make one more group of 3!

6. First draw one more marble in each picture. Then check if you can make one more group or not. Then write a division sentence.

a. 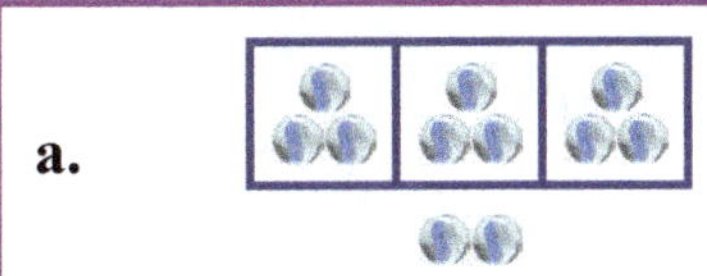	**b.** 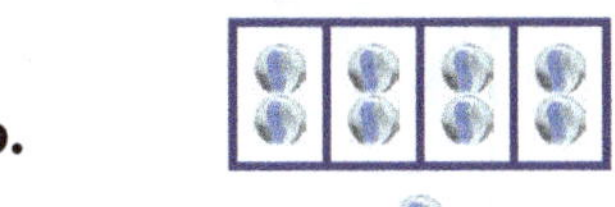	**c.** 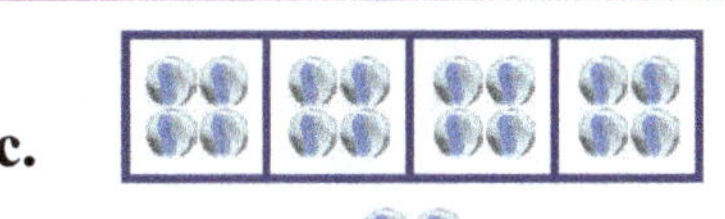
____ ÷ ____ = ____ R___	____ ÷ ____ = ____ R___	____ ÷ ____ = ____ R___

7. Solve, and find a pattern.

a. 21 ÷ 5 = ____ R ____	**b.** 56 ÷ 8 = ____ R ____	**c.** 43 ÷ 7 = ____ R ____
22 ÷ 5 = ____ R ____	57 ÷ 8 = ____ R ____	44 ÷ 7 = ____ R ____
23 ÷ 5 = ____ R ____	58 ÷ 8 = ____ R ____	45 ÷ 7 = ____ R ____
24 ÷ 5 = ____ R ____	59 ÷ 8 = ____ R ____	46 ÷ 7 = ____ R ____

8. Divide by 10. Indicate the remainder. Can you figure out a shortcut?

a. 29 ÷ 10 = ____ R ___	**b.** 78 ÷ 10 = ____ R ___	**c.** 54 ÷ 10 = ____ R ___
30 ÷ 10 = ____ R ___	79 ÷ 10 = ____ R ___	55 ÷ 10 = ____ R ___
31 ÷ 10 = ____ R ___	80 ÷ 10 = ____ R ___	56 ÷ 10 = ____ R ___

Puzzle Corner

The number sentence that *checks* the division is given. Write the corresponding division sentence.

a. 5 × 3 + 1 = 16	**b.** 7 × 4 + 3 = 31	**c.** 4 × 30 + 3 = 123
____ ÷ ____ = ____	____ ÷ ____ = ____	____ ÷ ____ = ____

Long Division 1

Divide hundreds, tens, and ones separately.

Write the dividend inside the long division "corner", and the quotient on top.

64 ÷ 2 = ?	**282 ÷ 2 = ?**
Divide tens and ones separately: 6 tens ÷ 2 = 3 tens (t) 4 ones ÷ 2 = 2 ones (o) 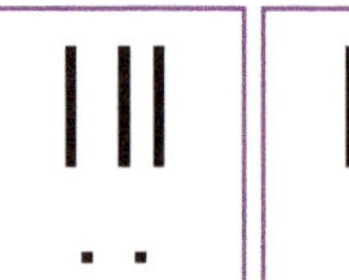 t o 3 2 2) 6 4	2 hundreds ÷ 2 = 1 hundred (h) 8 tens ÷ 2 = 4 tens (t) 2 ÷ 2 = 1 (o) 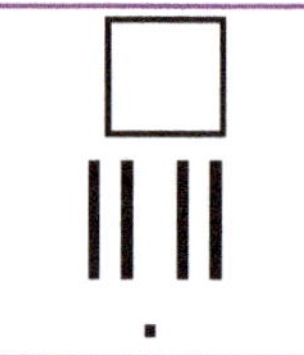 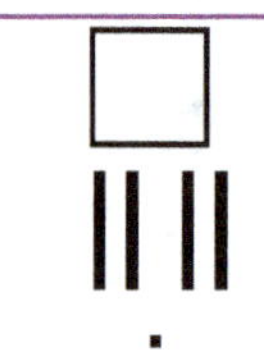 h t o 1 4 1 2) 2 8 2

1. Make groups. Divide. Write the dividend inside the "corner" if it is missing.

a. Make 2 groups	**b.** Make 3 groups	**c.** Make 3 groups	**d.** Make 4 groups
2)6 2	3)	3)	4)

2. Divide thousands, hundreds, tens, and ones separately.

a. 4)8 4 **b.** 3)3 9 3 **c.** 3)6 6 0 **d.** 4)8 0 4 0

e. 3)6 6 **f.** 2)6 0 4 2 **g.** 3)3 3 0 **h.** 4)4 8 0 4

h t o 0 4)2 4 8	h t o 0 6 2 4)2 4 8	th h t o 0 5)3 5 0 5	th h t o 0 7 0 1 5)3 5 0 5
Four does not go into 2. You can put zero in the quotient in the hundreds place or omit it. Four does go into 24, six times. Put 6 in the quotient.		Five does not go into 3. You can put zero in the quotient. Five does go into 35, seven times.	
Explanation: The 2 of 248 is 200 in reality. If you divided 200 by 4, the result would be less than 100, so that is why the quotient will not have any whole hundreds. Then you combine the 2 hundreds with the 4 tens. That makes 24 tens, and you CAN divide 24 tens by 4. The result, 6 tens goes as part of the quotient. Check the final answer: 4 × 62 = 248.		**Explanation:** 3,000 ÷ 5 will not give any whole thousands to the quotient because the answer is less than 1,000. But 3 thousands and 5 hundreds make 35 hundreds together. You can divide 3,500 ÷ 5 = 700, and place 7 as part of the quotient in the hundreds place. Check the final answer: 5 × 701 = 3,505.	

If the divisor does not "go into" the first digit of the dividend, look at the first two digits of the dividend.

3. Divide. Check your answer by multiplying the quotient and the divisor.

a. 3)1 2 3 (quotient: 0 4)

b. 4)2 8 4

c. 6)3 6 0

d. 8)2 4 8

e. 2)1 8 4

f. 7)4 2 7

g. 3)1 8 3 3

h. 4)2 4 0 4

i. 7)4 9 7 0

j. 5)4 5 0 5

The ones division is not even. There is a remainder.

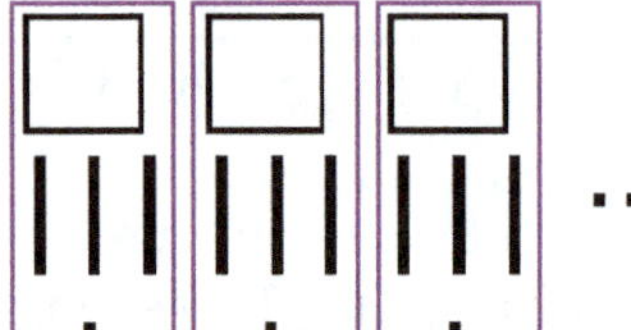

395 ÷ 3 = 131 R2

h	t	o
1	3	

3) 3 9 5

3 goes into 3 one time.
3 goes into 9 three times.

h	t	o	
1	3	1	R2

3) 3 9 5

3 goes into 5 one time, but not evenly.
Write the remainder 2 after the quotient.

h	t	o	
0	4	1	R1

4) 1 6 5

Four does not go into 1 (hundred). So combine the 1 hundred with the 6 tens (160).
Four goes into 16 four times.
Four goes into 5 once, with a remainder of 1.

th	h	t	o	
0	4	0	0	R7

8) 3 2 0 7

Eight does not go into 3 of the thousands. So combine the 3 thousands with the 2 hundreds (3,200).
Eight goes into 32 four times (3,200 ÷ 8 = 400)
Eight goes into 0 zero times (tens).
Eight goes into 7 zero times, with a remainder of 7.

4. Divide into groups. Find the remainder.

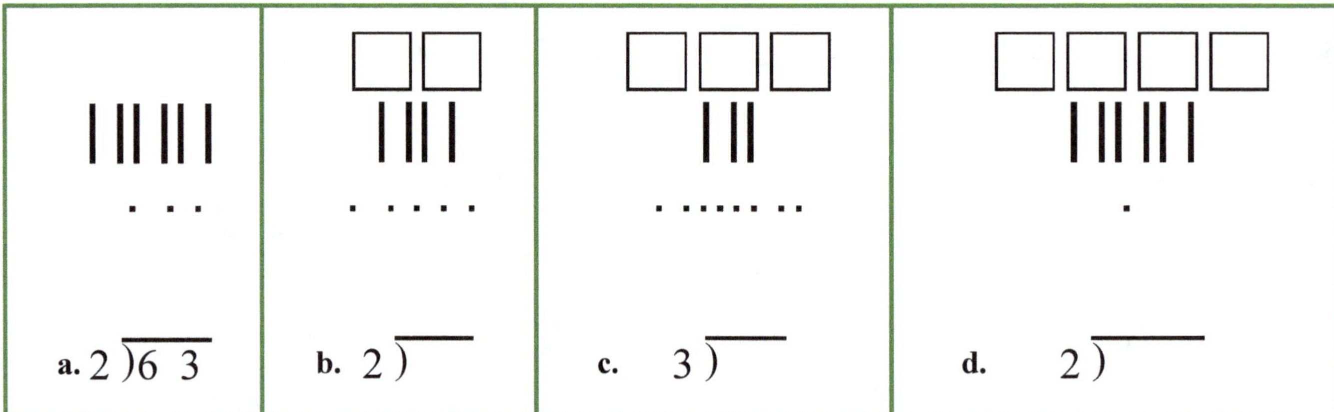

5. Divide. Indicate the remainder if any.

a. 4)8 4 7

b. 2)6 9

c. 3)3 6 7

d. 4)8 9

e. 2)1 2 1

f. 6)1 8 0 5

g. 7)2 1 5

h. 8)2 4 8 2

In the problems before, you just wrote down the remainder of the ones. Usually, we write down the subtraction that actually finds the remainder. Look carefully:

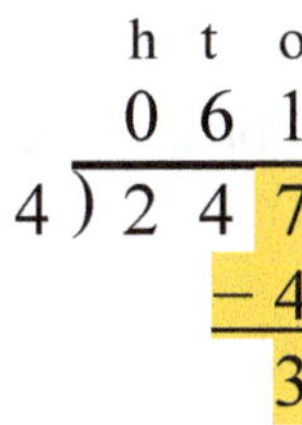

When dividing the ones, 4 goes into 7 one time. Multiply $1 \times 4 = 4$, write that four under the 7, and subtract. This finds us the remainder of 3.

Check: $4 \times 61 + 3 = 247$

```
   th h t o
    0 4 0 2
 4 ) 1 6 0 9
         - 8
           1
```

When dividing the ones, 4 goes into 9 two times. Multiply $2 \times 4 = 8$, write that eight under the 9, and subtract. This finds us the remainder of 1.

Check: $4 \times 402 + 1 = 1{,}609$

6. Practice some more. Subtract to find the remainder in the ones. Check your answer by multiplying the divisor times the quotient, and then adding the remainder. You should get the dividend.

a. 3)1 2 8

b. 3)9 5

c. 6)4 2 6 7

d. 4)2 8 4 5

e. 5)5 5 0 7

f. 2)8 0 6 3

7. Divide these numbers mentally. Remember, you can always check by multiplying!

a. $440 \div 4 =$	**b.** $3600 \div 400 =$	**c.** $824 \div 2 =$
$820 \div 2 =$	$369 \div 3 =$	$560 \div 90 =$

Long Division 2

Long division is a process of dividing into parts, starting from the biggest place value unit. For example, we divide the hundreds first, then the tens, then the ones. At each step, if we have a remainder, we combine that with the next unit we are going to divide.

> **Example 1.** Divide 78 by 3.
>
> First we divide the 7 tens by 3. That gives 2 tens for the quotient, and 1 ten left over that we couldn't divide. The 1 leftover ten is combined with the 8 ones. That is 18. Next, divide 18 by 3. That is 6 and there is no remainder. So, the division is over. The quotient is 2 tens and 6 ones, or 26. Check: $3 \times 26 = 78$.

If you could understand the above example, you will probably have no problem understanding the long division process as it is usually written out in the long division "corner". If not, don't worry just yet.

In long division, there are **three** processes going on in each step: 1) divide, 2) multiply and subtract to find the remainder, 3) combine the remainder with the next digit from the dividend.

1. Divide.	**2. Multiply and subtract.**	**3. Drop down the next digit.**
t o 2 2) 5 8	t o 2 2) 5 8 - 4 1	t o 2 9 2) 5 8 - 4 ↓ 1 8
Two goes into 5 two times, or 5 tens ÷ 2 = 2 whole tens—but there is a remainder!	To find it, multiply 2 × 2 = 4, write the 4 under the five, and subtract to find the remainder of 1 ten.	Next, drop down the 8 of the ones next to the leftover 1 ten. You combine the remainder ten with 8 ones, and get 18.

1. Divide.	**2. Multiply and subtract.**	**3. Drop down the next digit.**
t o 2 9 2) 5 8 - 4 1 8	t o 2 9 2) 5 8 - 4 1 8 - 1 8 0	t o 2 9 2) 5 8 - 4 1 8 - 1 8 0
Divide 2 into 18. Place 9 into the quotient.	Multiply 9 × 2 = 18, write that 18 under the 18, and subtract.	The division is over since there are no more digits in the dividend. The quotient is 29.

1. Divide.	2. Multiply and subtract.	3. Drop down the next digit.
t o 2 3) 8 4 Three goes into 8 two times, or 8 tens ÷ 3 = 2 whole tens—but there is a remainder!	t o 2 3) 8 4 - 6 2 To find it, multiply 2 × 3 = 6, write that 6 under the eight, and subtract to find the remainder of 2 tens.	t o 2 8 3) 8 4 - 4 ↓ 2 4 Next, drop down the 4 of the ones next to the leftover 2 tens. You combine the remainder tens with 4 ones, and get 24.

1. Divide.	2. Multiply and subtract.	3. Drop down the next digit.
t o 2 8 3) 8 4 - 6 2 4 Divide 3 into 24. Place 8 in the quotient.	t o 2 8 3) 8 4 - 6 2 4 - 2 4 0 Multiply 8 × 3 = 24, write that 24 under the 24, and subtract.	t o 2 8 3) 8 4 - 6 2 4 - 2 4 0 The division is over since there are no more digits in the dividend. The quotient is 28.

1. Divide.

a. 4) 6 4

b. 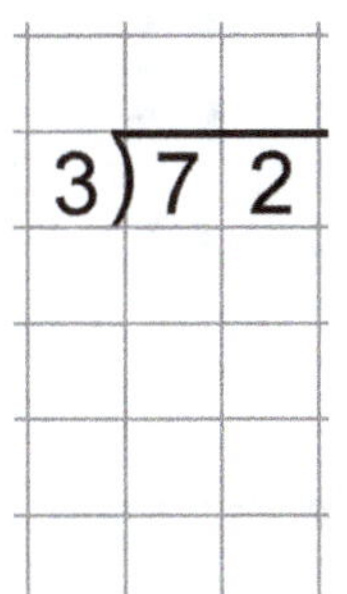

c. 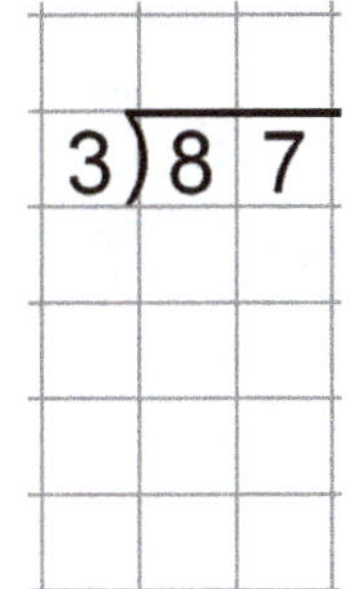

d. 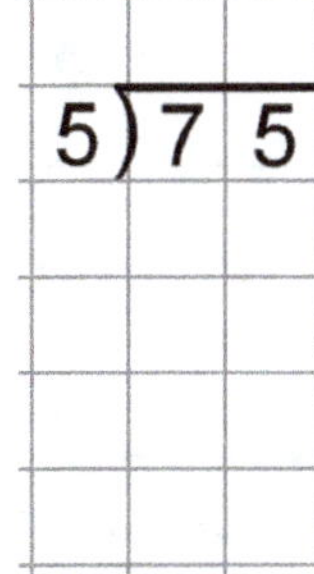

e. 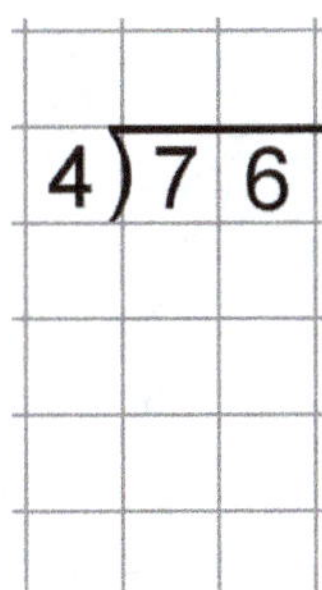

f. 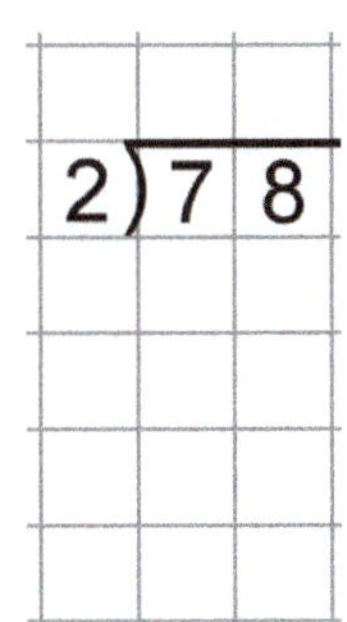

g. 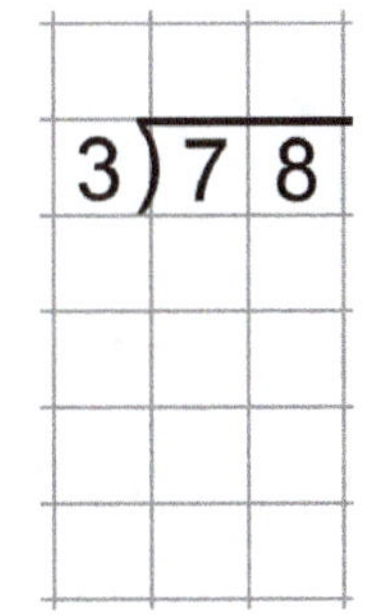

h. 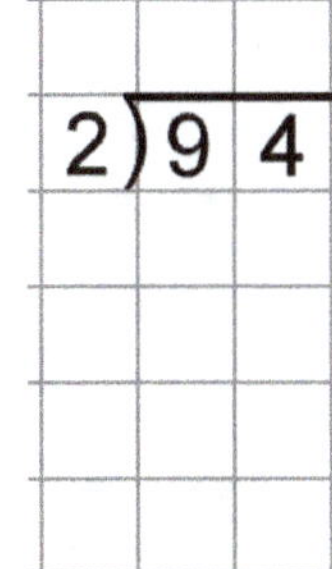

There are not enough hundreds, so look at two digits in the dividend.
You can place a zero in the quotient, or omit it.

h t o 0 5 4 $4\overline{)216}$ -2 0 1 6 - 1 6 0	h t o 0 6 5 $5\overline{)325}$ -3 0 2 5 - 2 5 0	h t o 0 3 8 $7\overline{)266}$ -2 1 5 6 - 5 6 0

2. Divide. Check each answer by multiplying.

a. 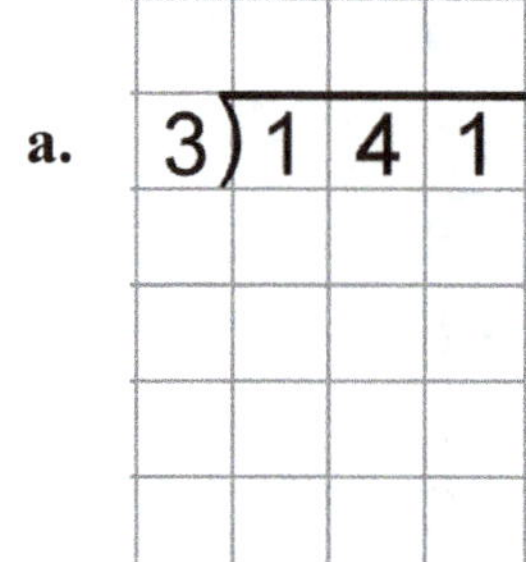

b. 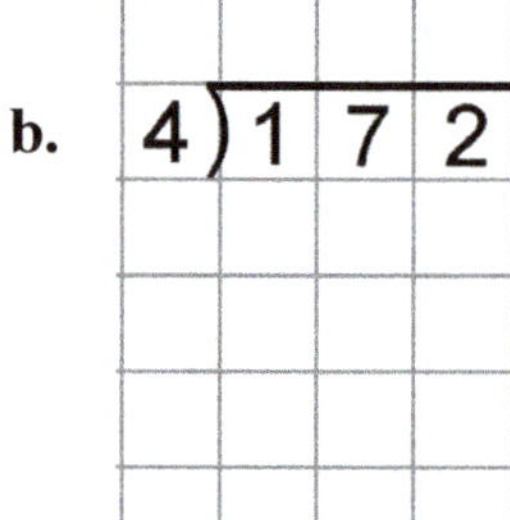

c. 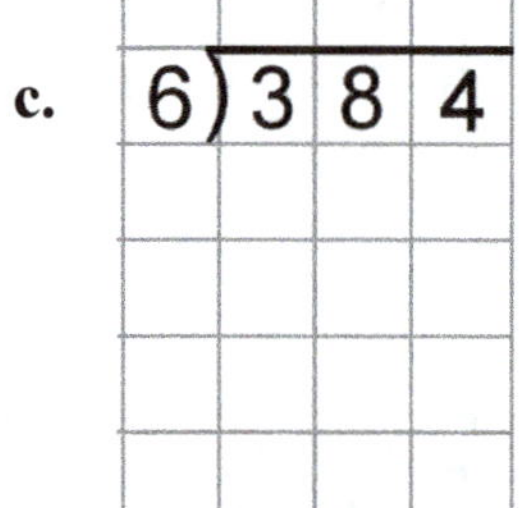

d. $8\overline{)272}$

e. $3\overline{)252}$

f. $7\overline{)406}$

Long Division 3

In this lesson we will divide three-digit numbers.

1. Divide.	2. Multiply and subtract.	3. Drop down the next digit.
h t o 1 2) 2 7 8 Two goes into 2 one time, or 2 hundreds ÷ 2 = 1 hundred.	h t o 1 2) 2 7 8 - 2 0 Multiply 1 × 2 = 2, write that 2 under the two, and subtract to find the remainder of zero.	h t o 1 2) 2 7 8 - 2 ↓ 0 7 Next, drop down the 7 of the tens next to the zero.
1. Divide.	**2. Multiply and subtract.**	**3. Drop down the next digit.**
h t o 1 3 2) 2 7 8 - 2 0 7 Divide 2 into 7. Place 3 into the quotient.	h t o 1 3 2) 2 7 8 - 2 0 7 - 6 1 Multiply 3 × 2 = 6, write the 6 under the 7, and subtract to find the remainder of 1 ten.	h t o 1 3 2) 2 7 8 - 2 0 7 - 6 1 8 Next, drop down the 8 of the ones next to the 1 leftover ten.
1. Divide.	**2. Multiply and subtract.**	**3. Drop down the next digit.**
h t o 1 3 9 2) 2 7 8 - 2 0 7 - 6 1 8 Divide 2 into 18. Place 9 into the quotient.	h t o 1 3 9 2) 2 7 8 - 2 0 7 - 6 1 8 - 1 8 0 Multiply 9 × 2 = 18, write that 18 under the 18, and subtract to find the remainder of zero.	h t o 1 3 9 2) 2 7 8 - 2 0 7 - 6 1 8 - 1 8 0 There are no more digits to drop down. The quotient is 139.

Can you follow these examples without the explanations?

Dividing the 8 hundreds.	Dividing the 25 tens.	Dividing the 12 ones.	Dividing the 7 hundreds.	Dividing the 17 tens.	Dividing the 21 ones.
1 6) 8 5 2 - 6 2	1 4 6) 8 5 2 - 6 2 5 -2 4 1	1 4 2 6) 8 5 2 - 6 2 5 -2 4 1 2 - 1 2 0	2 3) 7 7 1 - 6 1	2 5 3) 7 7 1 - 6 1 7 -1 5 2	2 5 7 3) 7 7 1 - 6 1 7 -1 5 2 1 - 2 1 0

1. Divide. Check each division with multiplication. The gridlines will help you keep your hundreds, tens, and ones lined up.

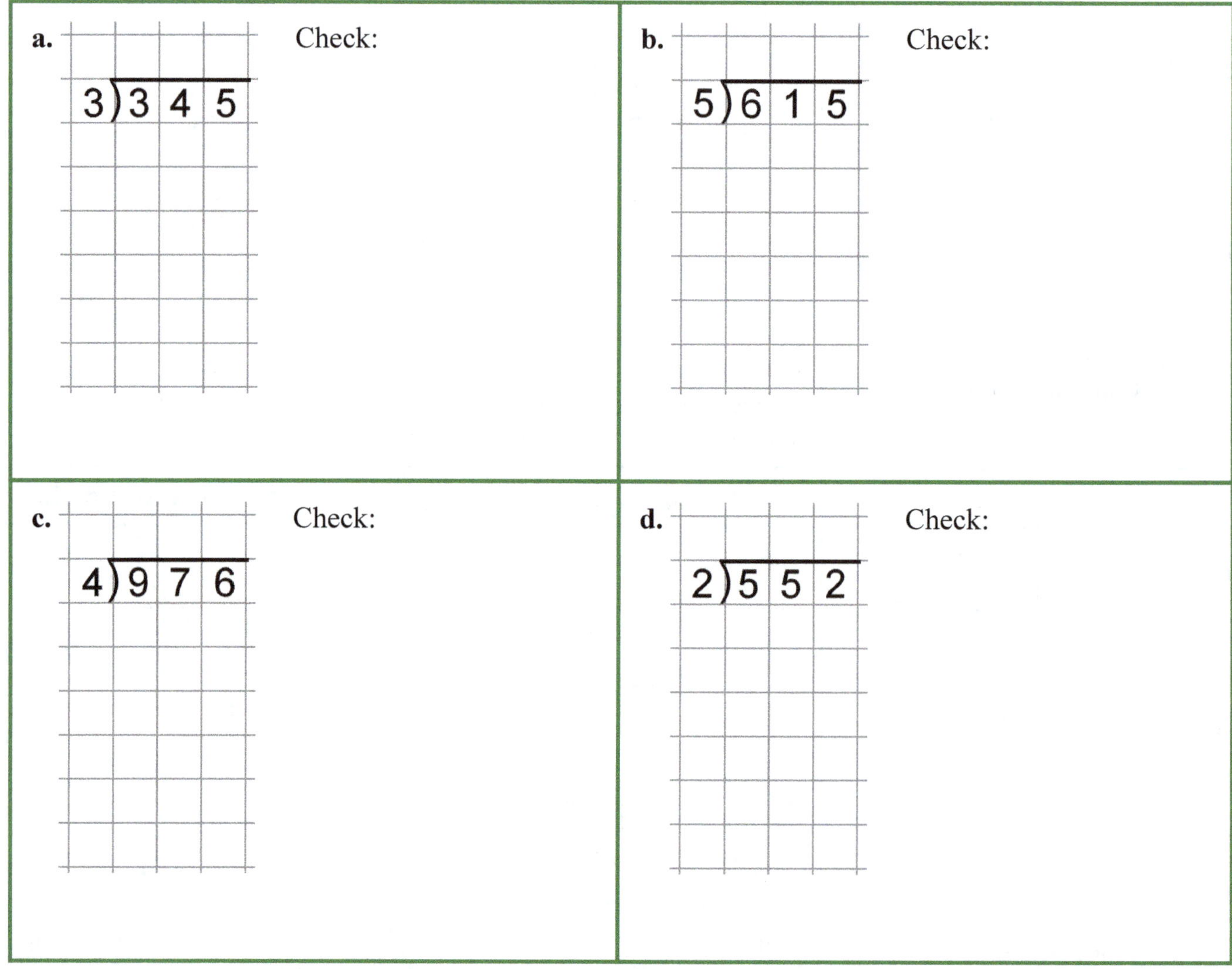

e. $3 \overline{)954}$ Check:	**f.** $7 \overline{)847}$ Check:
g. $4 \overline{)452}$ Check:	**h.** $5 \overline{)565}$ Check:
i. $2 \overline{)650}$ Check:	**j.** $7 \overline{)791}$ Check:
k. $8 \overline{)896}$ Check:	**l.** $4 \overline{)872}$ Check:

2. If you need more practice, then do these problems using the grids.
Check each one by multiplying.

a. 567 ÷ 3 Check:

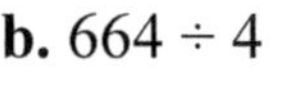

b. 664 ÷ 4 Check:

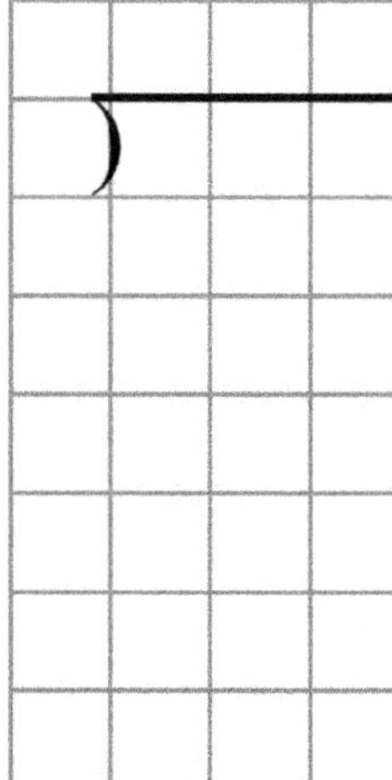

c. 994 ÷ 7 Check:

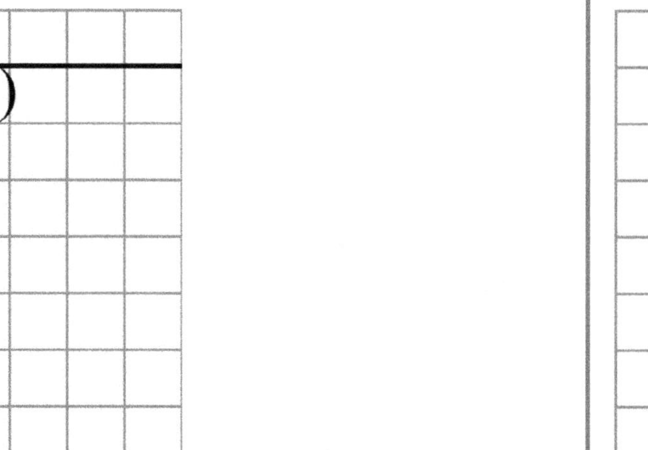

d. 585 ÷ 5 Check:

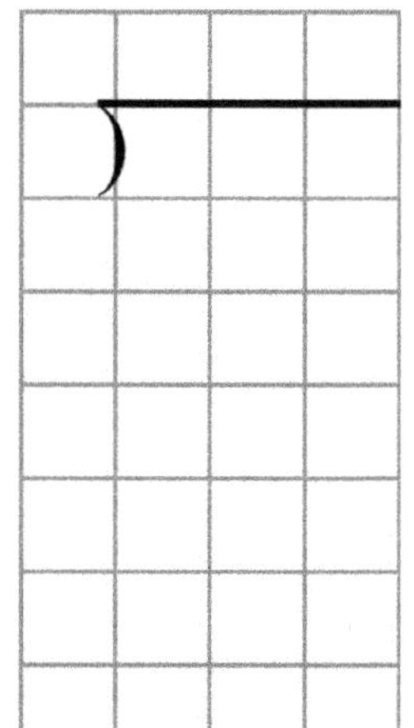

e. 912 ÷ 6 Check:

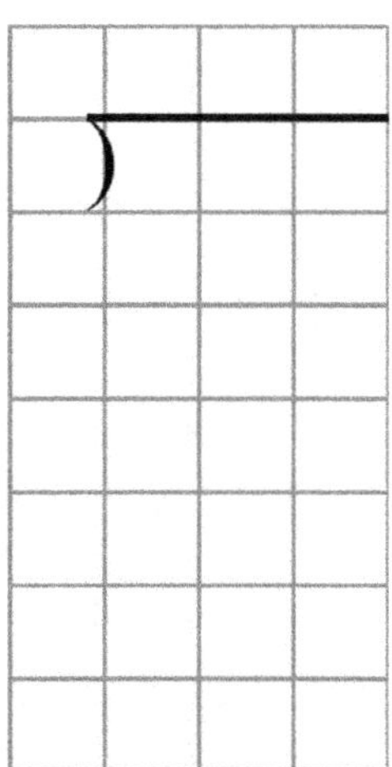

f. 936 ÷ 8 Check:

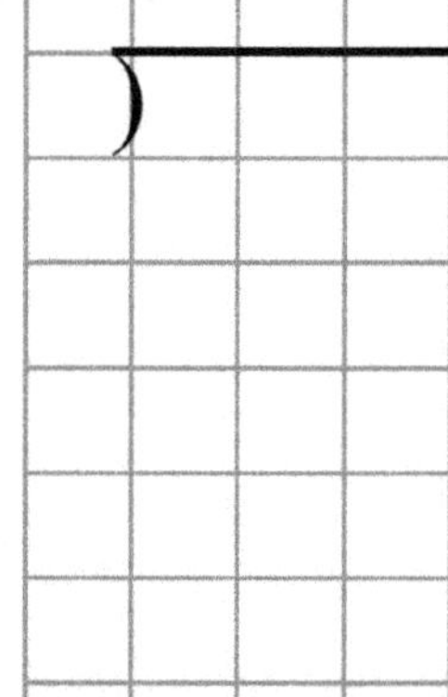

Long Division with 4-Digit Numbers

```
    th h  t  o        th h  t  o        th h  t  o        th h  t  o
     2                 2  8              2  8  5           2  8  5  9
 2 ) 5  7  1  8    2 ) 5  7  1  8    2 ) 5  7  1  8    2 ) 5  7  1  8
   - 4               - 4               - 4               - 4
     1                 1  7              1  7              1  7
                     - 1  6            - 1  6            - 1  6
                          1                 1  1              1  1
                                          - 1  0            - 1  0
                                                1                1  8
                                                               - 1  8
                                                                    0
```

Long division with 4-digit numbers works the same way as with smaller numbers!

Check:

$$\begin{array}{r} 2\,8\,5\,9 \\ \times \quad\;\; 2 \\ \hline \end{array}$$

1. Divide. Check each division result with multiplication.

a. $3\overline{)7041}$  Check:

b. $4\overline{)9240}$ Check:

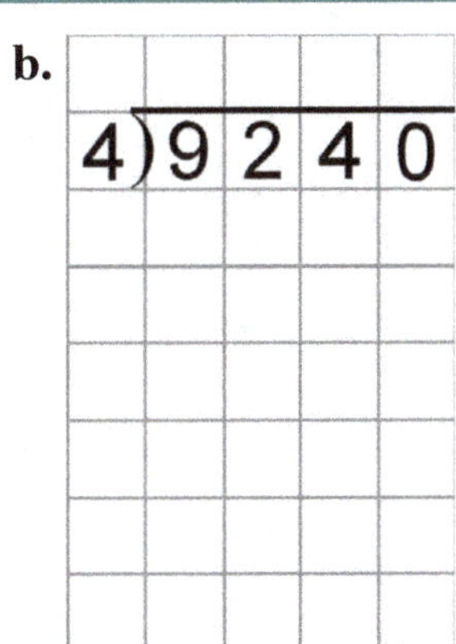

c. $4\overline{)7140}$ Check:

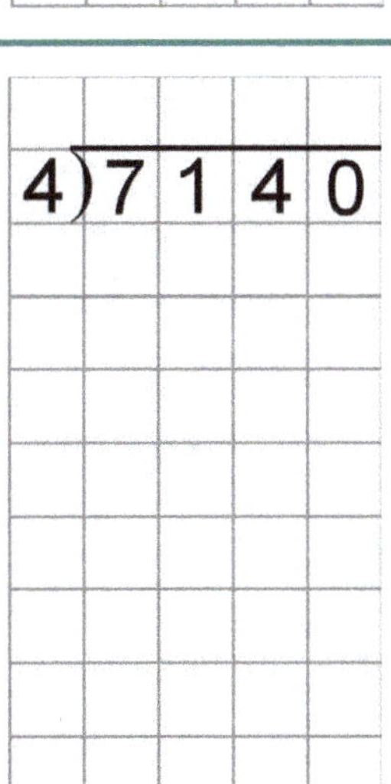

d. $2\overline{)9770}$ Check:

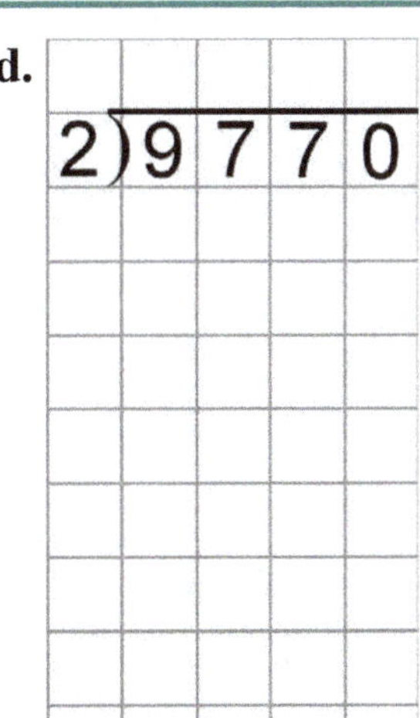

2. Divide. Use the grids below. Check each one by multiplication.

a. 5802 ÷ 3 Check: 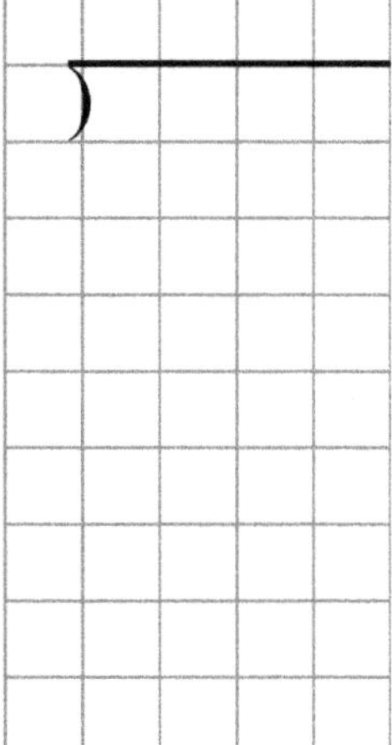	**b.** 1653 ÷ 3 Check: 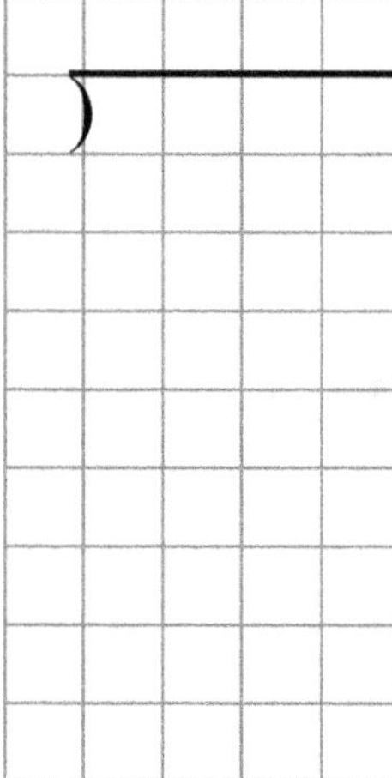
c. 9380 ÷ 7 Check: 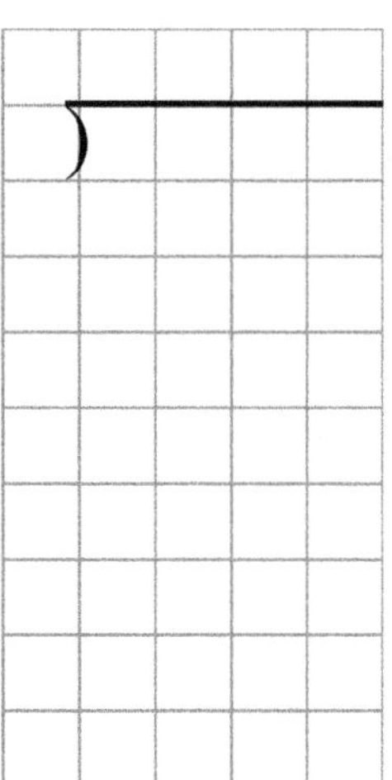	**d.** 9104 ÷ 8 Check: 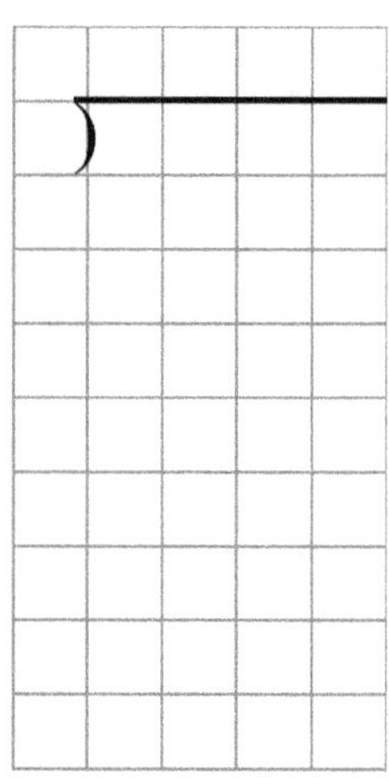
e. 7902 ÷ 6 Check: 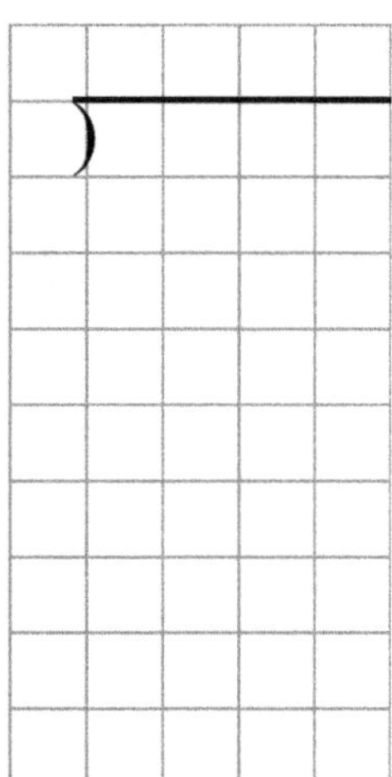	**f.** 6080 ÷ 5 Check: 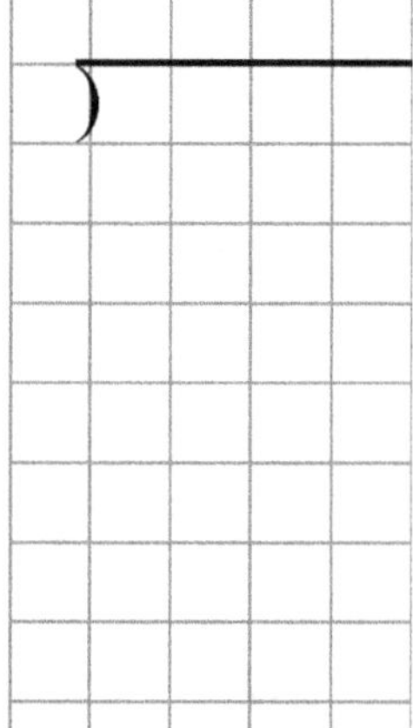

There are not enough thousands. So, when you start, look at the **first two digits** of the dividend, and divide the divisor into those.	$\begin{array}{r} 04 \\ 7\overline{)3052} \\ \underline{-28} \\ 25 \end{array}$ 7 does not go into 3, so look at the **first two digits** ("30"). 7 goes into 30 four times.	$\begin{array}{r} 043 \\ 7\overline{)3052} \\ \underline{-28} \\ 25 \\ \underline{-21} \\ 42 \end{array}$ 7 goes into 25 three times. 4 tens is the remainder.	$\begin{array}{r} 0436 \\ 7\overline{)3052} \\ \underline{-28} \\ 25 \\ \underline{-21} \\ 42 \\ \underline{-42} \\ 0 \end{array}$ 7 goes into 42 six times.

3. Divide. You may need to look at the first two digits of the dividend. Check your answers.

a. $3\overline{)1479}$ Check:	**b.** $5\overline{)1920}$ Check:
c. $6\overline{)5544}$ Check:	**d.** $5\overline{)245}$ Check:
e. $6\overline{)522}$ Check:	**f.** $9\overline{)3339}$ Check:

4. Two neighbors bought nine trees for $16 each, and shared the cost equally. How much did each person pay?

5. A DVD contains a total of 504 minutes of show episodes. The Mayors decided to watch them all in a week, and to watch the same amount each day.

   How many minutes will they watch each day?

   Express this also as hours and minutes.

6. A path through the woods is 2,600 feet long and it is divided into eight parts of equal length. Clues for a treasure hunt are placed at those points. The children get their first clue at the beginning of the path.

   a. At what distance from the start is the second clue?

   b. What is the distance to the third clue?

7. Cindy bought eight bags of balloons for her daughter's birthday party, because they were on sale. Each bag had 25 balloons. She took 96 balloons out and gave six balloons to each child.

   a. How many children were going to be at the party?

   b. How many balloons does Cindy have left?

More Long Division

<table>
<tr>
<td rowspan="2">Study the example carefully. We need to place a zero in the quotient.</td>
<td>
<pre>
    2 4
4)9 6 2 0
  -8
   1 6
</pre>
</td>
<td>
<pre>
    2 4 0
4)9 6 2 0
  -8
   1 6
  -1 6
     0 2
</pre>
</td>
<td>
<pre>
    2 4 0 5
4)9 6 2 0
  -8
   1 6
  -1 6
     0 2 0
     - 2 0
         0
</pre>
</td>
</tr>
<tr>
<td>At this point, the division is even. Continue normally, multiplying 4 × 4 = 16.</td>
<td>Now, 4 does not go into 02, so place a zero in the quotient. You can either continue as usual, or drop another digit from the dividend.</td>
<td>Dropping another digit from the dividend, we get 020. 4 goes into 20 five times.</td>
</tr>
</table>

1. Let's practice! There will be a zero in the quotient. Multiply to check.

a. 5)5 2 2 5 Check:

b. 3)4 2 1 8 Check:

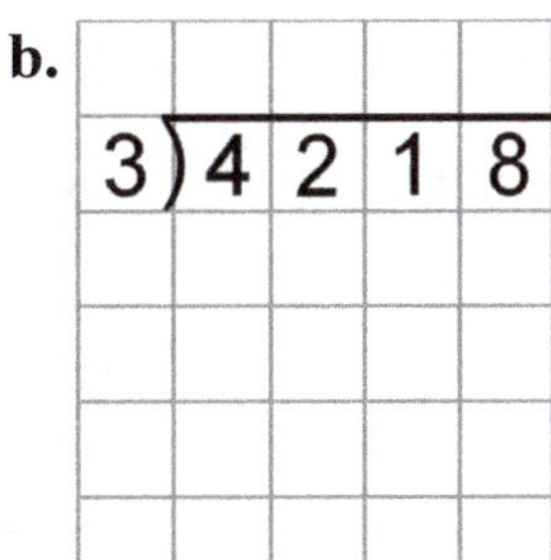

c. 4)8 1 4 8 Check:

d. 7)9 1 4 9 Check:

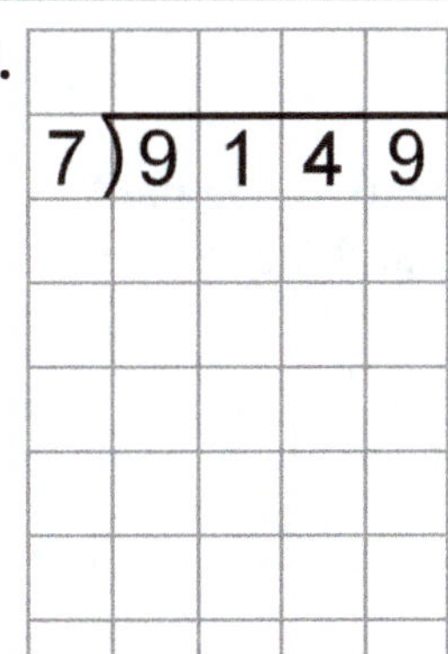

2. For more practice, do these in a notebook or blank paper. Use grid paper if possible.

a. 8,115 ÷ 3 **b.** 6,540 ÷ 5 **c.** 9,163 ÷ 7 **d.** 6,378 ÷ 6

Short, even division			
2,156 ÷ 7 is easy to do mentally: 2,100 ÷ 7 is 300, and 56 ÷ 7 is 8. So the answer is 308. If we use long division, we get a really "short" division. That is because the division is even in the first two digits and in the last two digits. So, you don't have to multiply and subtract.	th h t o 0 3 7) 2 1 5 6 7 goes into 21 three times.	th h t o 0 3 0 7) 2 1 5 6 7 goes into 5 zero times...	th h t o 0 3 0 8 7) 2 1 5 6 ...so look at the two digits (56). 7 goes into 56 eight times.

3. Divide.

a. 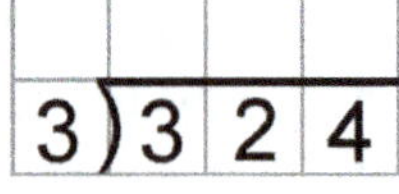
3) 3 2 4

b. 
4) 8 2 0

c. 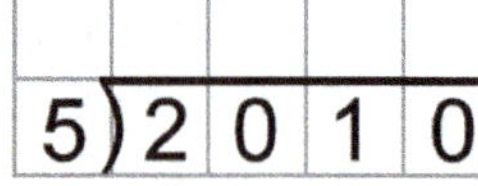
5) 2 0 1 0

d. 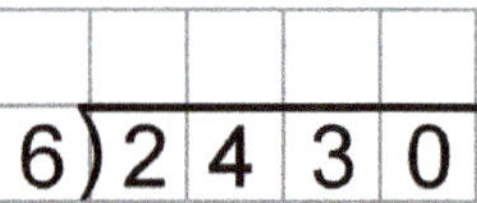
6) 2 4 3 0

e. 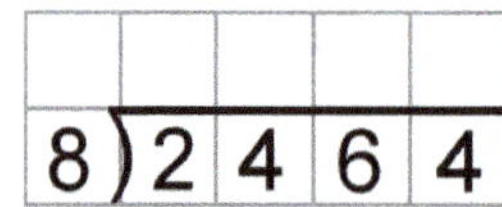
8) 2 4 6 4

f. 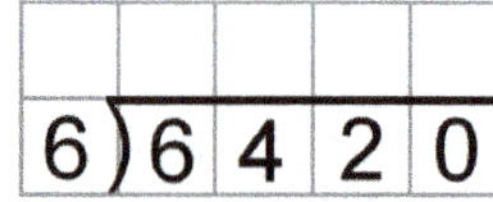
6) 6 4 2 0

4. Mary divided 285 buttons evenly into the five compartments. Find out how many buttons are

a. in one compartment.

b. in three compartments.

c. in four compartments.

5. Dad bought a used car for $9,620. He paid $2,000 as a down payment, and the rest was paid in four equal payments. How much was each payment?

(This page is optional.)

6. Here you can try your division skills with a 5-digit dividend.

The process is the same, just longer, since the dividend has more digits.

Lastly, check with multiplication, as you should always do.

a. $3\overline{)63702}$ Check:

b. $2\overline{)70814}$ Check:

c. $2\overline{)43290}$ Check:

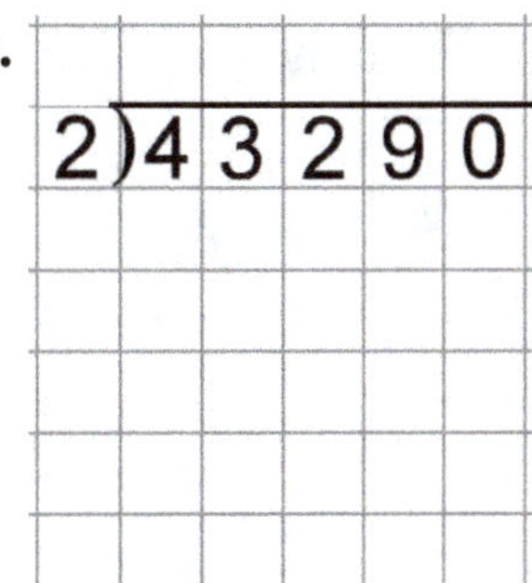

d. $5\overline{)15810}$ Check:

e. $9\overline{)47475}$ Check:

Remainder Problems

When using long division, the division is not always exact.

Example 1. At this point there are no more digits to drop down from the dividend. The last subtraction yields 6, which is the remainder. So 125 ÷ 7 = 17 R6.

Note that the remainder 6 is LESS THAN 7, the divisor.

```
     1 7
   ______
 7 ) 1 2 5
   - 7
   ---
     5 5
   - 4 9
   -----
       6
```

Check: Multiply the answer (17) by the divisor (7).

Then add the remainder (6). You get the original dividend (125).

```
    4
    1 7
  ×   7
  -----
  1 1 9
+     6
  -----
  1 2 5
```

1. Divide. Check each result in the empty space by multiplication and addition.

a. 514 ÷ 3 Check:

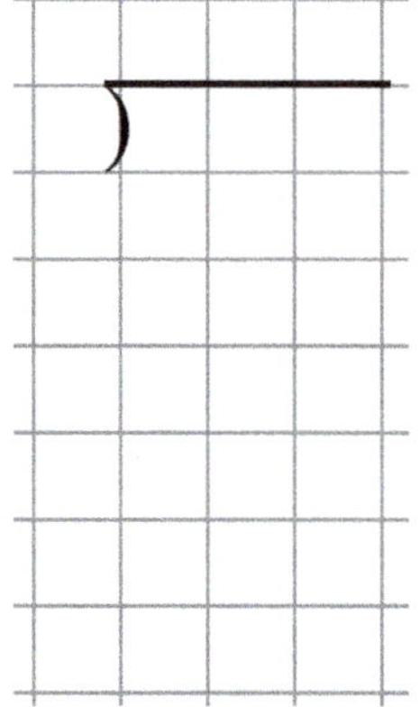

b. 673 ÷ 8 Check:

c. 1,905 ÷ 6 Check:

d. 8,205 ÷ 4 Check:

2. Find the divisions that are incorrect. Redo the ones that are wrong below.

a.

77
6) 4 6 3
-4 2
4 9
- 4 2
7

b.

3 5 3
7) 2 4 7 3
-2 1
3 7
- 3 5
2 3
- 2 1
2

c.

3 5 1
9) 4 0 5 9
-3 6
4 5
- 4 5
0 9

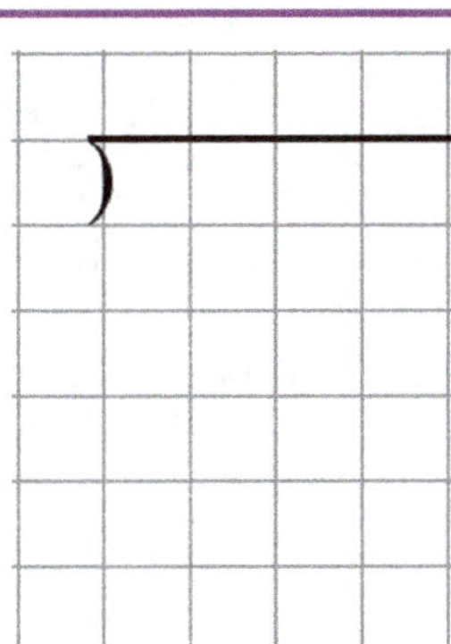

d. How can you spot the error in (a) just by looking at the remainder 463 ÷ 6 = 77 **R7**?

3. Write a division sentence for each problem and solve it. Lastly, explain what the answer means.

a. Arrange 112 chairs into rows of 9.

____

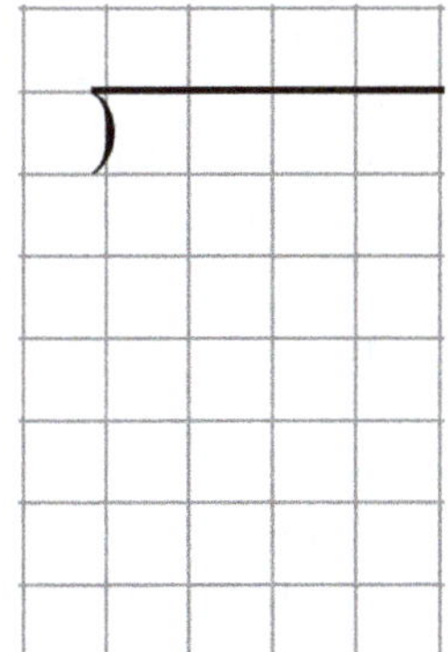

We get _______ rows, _______ chairs in each row, and ________________________________

____

b. Arrange 800 erasers into piles of 3.

____

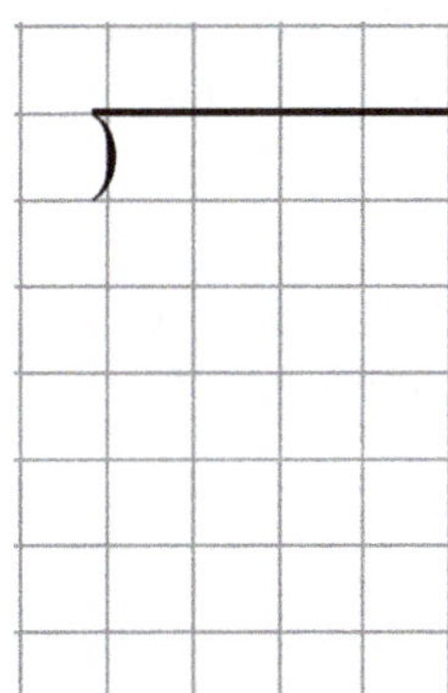

We get _______ piles, _______ erasers in each pile, and ________________________________

____

Imagine you are trying to pack things evenly into some "containers," and they don't go evenly. The last container will not be full. Often students make mistakes with such problems. Read the question carefully. Sometimes you *do* need to count the container that is not full, and sometimes not.	
Example 2. One hundred forty-six people were transported in vans that could carry 9 passengers each. How many vans were needed? The division is 146 ÷ 9 = 16 R2, which means that 16 vans were full and one van had 2 passengers. But the answer is that they needed **17** vans!	**Example 3.** You have 1,250 blank CDs. You pack 200 into each box. How many *full* boxes will there be? 6 × 200 = 1,200, from which we get that 1,250 ÷ 200 = 6 R50. Six boxes will be full (and 50 CDs are left over, perhaps not even packed).

Solve the problems. Sometimes you can use *multiplication* instead of division.

4. A company bags 2,000 lb of potatoes into 12-lb bags.
   The division is: 2,000 ÷ 12 = 166 R8.
   How many full bags will they get?

5. If 50 people can fit into one bus, how many buses would you need to transport 940 people?

6. Mr. Eriksson can have 75 days of vacation each year.
   He wants to divide those days into 4 vacations.
   What will the length of each vacation be?
   Make them as close to the same length as possible.

7. A farm packs 400 kg of strawberries so that they make ninety 2 kg boxes, forty 4 kg boxes, and the rest is packed into 6 kg boxes. How many *full* 6 kg boxes will they get?

8. Can you pack 412 tennis balls into containers evenly so that each container has...

   a. 4 balls?

   b. 5 balls?

   c. 6 balls?

9. Mr. Sanders needs to paint 740 wooden pieces, using 6 different colors—red, orange, yellow, green, blue, and purple—in *nearly* equal amounts. How many pieces will there be of each color?

10. Do **one problem** from each box by long division. Can you then figure out the answers to the other two in each box, without actually dividing?

a. 211 ÷ 3 = 212 ÷ 3 = 213 ÷ 3 =	**b.** 1,206 ÷ 7 = 1,207 ÷ 7 = 1,208 ÷ 7 =
c. 411 ÷ 5 = 412 ÷ 5 = 413 ÷ 5 =	**d.** 7,185 ÷ 9 = 7,186 ÷ 9 = 7,187 ÷ 9 =

11. *A challenge:* if 231 ÷ 6 = 38 R3, then figure out what 232 ÷ 6 is.

12. Divide these numbers by 10 and indicate the remainder. There is a shortcut!

a. 787 ÷ 10 = 66 ÷ 10 = 340 ÷ 10 =	**b.** 452 ÷ 10 = 509 ÷ 10 = 52 ÷ 10 =	**c.** 463 ÷ 10 = 982 ÷ 10 = 925 ÷ 10 =

Puzzle Corner

What is wrong with the division result 31 ÷ 6 = 4 R7? After all, 4 × 6 + 7 = 31, so it seems to check fine. Explain.

Long Division with Money

Long division with money amounts is done the same way as with whole numbers.
We just place a decimal point in the quotient in the same place as where it is in the dividend.

Complete the divisions below, and check them with multiplication.
After completing the problems 1 a and 1 b, check with your teacher whether you can continue.

1. Divide and check with multiplication.

a. \$25.41 ÷ 3 Check:

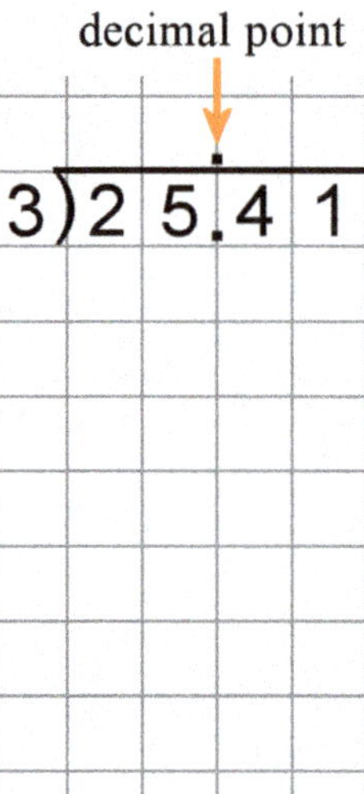

b. \$14.88 ÷ 4 Check:

2. Solve the problems.

a. Amy, Sally, and Joe shared equally a salary of \$85.50. How much did each one get?

b. If a gallon of milk costs \$4.56, what would one quart cost?

Example 1. A pizza costs \$25.55. If four people share the cost evenly, how much does each person have to pay?

The division result is \$25.55 ÷ 4 = \$6.38 R3¢. The division is not even.

Each person's share is \$6.38. There is an extra 3¢, so in reality three of the people would pay \$6.39, and one person would pay \$6.38.

```
     6.3 8
4 ) 2 5.5 5
   - 2 4
     ---
       1 5
     - 1 2
       ---
         3 5
       - 3 2
         ---
           3
```

3. Margie and Annie bought a handbag for \$25.56, a mirror for \$3.55, and a brush for \$2.75. They shared the cost equally. How much did each girl pay?

4. Three people shared the cost of movie tickets that cost \$25.95 and some popcorn for \$4.35. How much was each person's share?

5. Joe bought a camera for \$358.60. His dad paid \$100 of it for him, and the rest of the payment was divided into four equal monthly payments. How much was each payment?

6. A gallon of ice cream costs \$12.96. You and your brother each pay for one-eighth part of the cost, and Mom will pay the rest. Find each person's share of the cost. (*Hint: to find one-eighth part of something, divide by 8.*)

You: _____________ Your brother: _____________

Mom: _____________

Long Division Crossword Puzzle

1. Divide. Place each answer in the cross-number puzzle. Use your notebook or the grid below.

Across:

a. 3,440 ÷ 8

b. 574 ÷ 7

c. 234 ÷ 9

d. 1,707 ÷ 3

e. 4,756 ÷ 2

Down:

a. 1,072 ÷ 8

b. 6,135 ÷ 3

c. 145 ÷ 5

d. 2,652 ÷ 4

e. 1,442 ÷ 7

f. 3,474 ÷ 9

a.					
		b.		b.	e.
a.					
				c.	
		d.	d.		
					f.
		e.			

Average

The Millers went on a trip. The first day, they drove 110 miles, the second day, 142 miles, the third day, 126 miles, and the last day, 82 miles. The Millers drove a total of 460 miles.

In the diagram, we have put those distances as sticks one after another, though of course in reality they did not drive just straight stretches of roads.

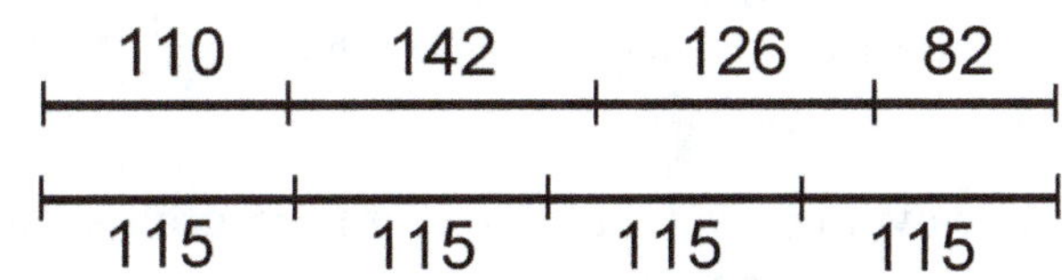

IF they had driven 115 miles each day, it would have totaled the same 460 miles.

On average, the Millers drove 115 miles a day, or their ***average*** daily mileage was 115 miles.

What is the average of 20, 32, 27, 37, and 24?
First find the total by adding. Then, divide that into equal parts.

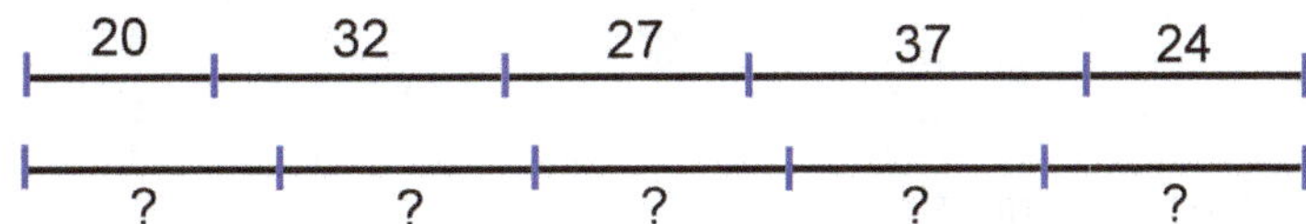

$20 + 32 + 27 + 37 + 24 = 140$. $140 \div 5 = 28$. So, the **average** of 20, 32, 27, 37, and 24 is 28.

If these numbers were the ages of club members, we would say the average age of the members is 28 years. However, they could also be distances, weights, volumes, or just plain numbers.

1. Judith's test scores were 78, 87, 69, and 86.
   Find her average score.

2. John measured the temperature five times during a day.
   These are the results that he recorded:
   18°C, 22°C, 26°C, 23°C, and 16°C.
   Find the average temperature for the day.

3. Dad drove 414 km in six hours.
   How many kilometers did he drive, on the average, in one hour?

You can also figure the average backwards.
During a 20-hour drive from Denver to Dallas, Dad's average speed was 40 miles per hour. How far is Denver from Dallas?   You can multiply to find the answer: 20 hours × 40 miles/hour = 800 miles.   (Note that in reality, he did not drive with a constant speed all of the time because he had to stop at crossings, slow down on curves, stop for a snack and so on. We do not know how much his speed varied on the trip. All we are given is that his *average* speed was 40 miles per hour. Of course the average speed was calculated by dividing the length of the trip by the total number of hours the trip took.)

4. The average weight of an egg is 55 grams. How much would a dozen eggs weigh?

5. For her hospital stay, Mom was charged an average of \$76 daily. What was the total cost of her one-week stay?

6. Mom's weekly grocery bills in June were \$234, \$178, \$250, and \$198. What was her average weekly grocery bill?

7. The children ran a race. These are the resulting times:

Ann	12 min
Judy	15 min
Rose	14 min
Elizabeth	19 min
Grace	12 min
Nancy	18 min

Michael	12 min
Greg	10 min
James	11 min
Caleb	15 min
Hans	17 min

Find the girls' average running time and the boys' average running time <u>*separately*</u>.

Girls' average: _______________

Boys' average: _______________

Are boys or girls quicker on average?

What is the difference of the two averages?

8. Here are the science quiz scores for ten fourth-graders: 24 20 24 16 28 30 14 22 23 19

a. Finish the frequency table and the graph. ("Frequency" refers to the number of students.)

Quiz score	Frequency
13-15	1
16-18	1
19-21	2
22-24	
25-27	
28-30	

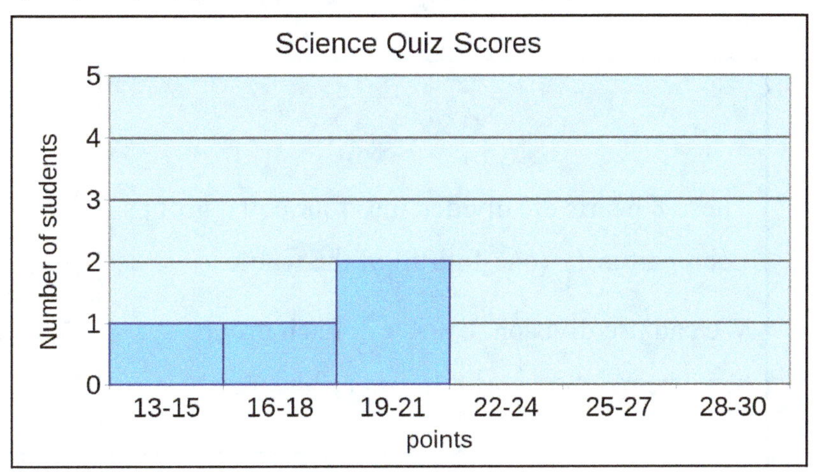

b. Calculate the average score.

c. Both the graph and the average tell us what the "middle" or "typical" result in the test was. Explain how you can guess what the average is approximately, just using the graph.

9. These are the ages of the members of a bird watching club:

18 28 25 33 29 17 44 37 30

a. Calculate the average age.

b. The club gained a new member, 79-year-old Jim. What is the average age now?

Puzzle Corner

If 213 ÷ 17 = 12 R9, what is 213 ÷ 12?

Finding Fractional Parts with Division

 These 8 hearts are divided into four equal groups. Each part is $\frac{1}{4}$ (one-fourth) of the whole. We can use division: $8 \div 4 = 2$. Each group has 2 hearts. So, $\frac{1}{4}$ of 8 hearts is 2 hearts.	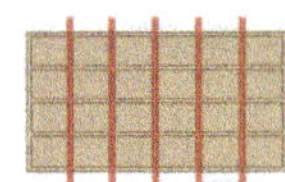 Mom divided 24 brownies into 6 equal parts. Each part is 1/6th of the whole. How many pieces are in each part? Divide to find out: $24 \div 6 = 4$. Four pieces. So, $\frac{1}{6}$ of 24 brownies is 4 brownies.

To find $\frac{1}{2}$, $\frac{1}{3}$, $\frac{1}{4}$, $\frac{1}{5}$ *etc.* part of something, divide by 2, 3, 4, 5, *etc.* (respectively).

1. Write a division sentence and a fractional part sentence.

a. 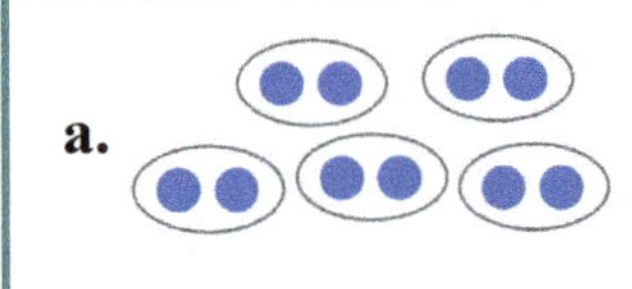	**b.**	**c.** 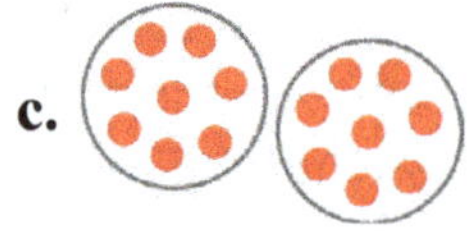	**d.** 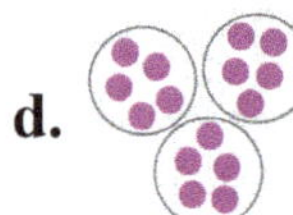
____ ÷ 5 = ____	____ ÷ ___ = ___	____ ÷ ___ = ___	____ ÷ ___ = ___
$\frac{1}{5}$ of ____ is ____.	$\frac{1}{3}$ of ____ is ____.	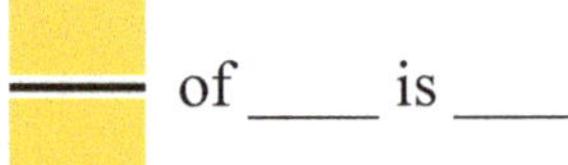 of ____ is ____.	of ____ is ____.

2. Write a fractional part sentence for each division sentence.

a. $30 \div 5 =$ _____	**b.** $48 \div 6 =$ _____	**c.** $25 \div 5 =$ _____	**d.** $50 \div 5 =$ _____
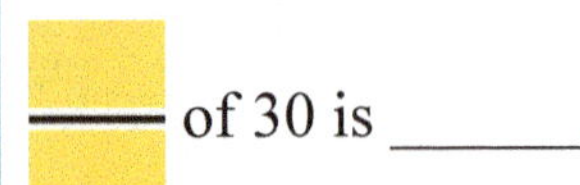 of 30 is _____.	of ____ is ____.	of ____ is ____.	of ____ is ____.

3. Find a part. Also write a division sentence.

a. $\frac{1}{6}$ of 30 is ______.	**b.** $\frac{1}{7}$ of 49 is ______.	**c.** $\frac{1}{10}$ of 250 is ______.
30 ÷ 6 = 5	______ ÷ ____ = ____	______ ÷ ____ = ____
d. $\frac{1}{2}$ of 480 is ______.	**e.** $\frac{1}{9}$ of 1,800 is ______.	**f.** $\frac{1}{5}$ of 400 is ______.
______ ÷ ____ = ____	______ ÷ ____ = ____	______ ÷ ____ = ____

Divide these ten fish into 5 groups.

1/5 of 10 fish is 2 fish. $\frac{1}{5}$ of 10 is 2.

2/5 of 10 fish is twice as much, or 4 fish. $\frac{2}{5}$ of 10 is 4.

3/5 of 10 fish is three times as much, or 6 fish. $\frac{3}{5}$ of 10 is 6.

What about 4/5 of 10? Can you tell? $\frac{4}{5}$ of 10 is _____.

4. Fill in the blanks.

a.

$\frac{1}{3}$ of 9 apples is _____ apples.

$\frac{2}{3}$ of 9 apples is _____ apples.

$\frac{3}{3}$ of 9 apples is _____ apples.

b.

$\frac{1}{4}$ of 12 flowers is _____.

$\frac{2}{4}$ of 12 flowers is _____.

$\frac{3}{4}$ of 12 flowers is _____.

$\frac{4}{4}$ of 12 flowers is _____.

c.

$\frac{1}{5}$ of _____ fish is _____ fish.

$\frac{2}{5}$ of _____ fish is _____ fish.

$\frac{3}{5}$ of _____ fish is _____ fish.

$\frac{4}{5}$ of _____ fish is _____ fish.

$\frac{5}{5}$ of _____ fish is _____ fish.

5. Calculate.

a. $\frac{1}{5}$ of 20 is _______.

$\frac{2}{5}$ of 20 is _______.

$\frac{3}{5}$ of 20 is _______.

b. $\frac{1}{8}$ of 32 is _______.

$\frac{3}{8}$ of 32 is _______.

$\frac{5}{8}$ of 32 is _______.

c. $\frac{1}{10}$ of 500 is __________.

$\frac{3}{10}$ of 500 is __________.

$\frac{7}{10}$ of 500 is __________.

d. $\frac{1}{6}$ of 420 is __________.

$\frac{2}{6}$ of 420 is __________.

$\frac{5}{6}$ of 420 is __________.

e. $\frac{1}{20}$ of 600 is __________.

$\frac{7}{20}$ of 600 is __________.

$\frac{11}{20}$ of 600 is __________.

f. $\frac{1}{100}$ of 700 is __________.

$\frac{9}{100}$ of 700 is __________.

$\frac{50}{100}$ of 700 is __________.

6. Fill in.

a. Marsha's mom gave her $18. She put $6 into her savings, which was one-__________ part of it.

Division sentence: ________ ÷ _____ = ______

b. Mariana had $80 in savings and spent one-fourth of it, or $________.

Division sentence: ________ ÷ _____ = ______

c. One-______ of the group of 25 boys went jogging, which meant 5 boys went jogging.

Division sentence: ________ ÷ _____ = ______

7. An 8-lb bag of potatoes costs $3.20.

a. One pound is one-__________________ part of the bag. How much does it cost?

b. How much does 5/8 of the bag cost?

8. A shepherd's pie is divided into 10 equal-size pieces. The whole pie weighs 1,200 g.

a. What does 1/10 of the pie weigh?

b. What does 9/10 of the pie weigh?

9. Four backpacks cost a total of $28. How much would three backpacks cost?

10. A new fan for the clubhouse costs $24.40. Mark paid for half of it, Judy paid ¼ of it, and Art and Grace each paid ⅛ of it. Find how much each child paid.

Mark: ______________ Judy: ______________

Art: ______________ Grace: ______________

11. Erica and James had 28 balloons each to sell.

By the evening, Erica had sold ½ of her balloons.
James had sold ¾ of his.
How many balloons had they sold *in total*?
Hint: First calculate how many balloons Erica sold.

Problems with Fractional Parts

1. A loaf of bread weighs 400 g. It is cut into 20 slices.

   a. How much does one slice weigh?

   b. How much do three slices weigh?

   c. How much do 11 slices weigh?

2. A pie weighs 450 g. It is cut into six equal-size pieces.

   a. How much does 2/6 of the pie weigh?

   b. How much does 5/6 of the pie weigh?

3. If you need to calculate 5/9 of the number 729,

   first divide 729 by ______,

   then ____________________ the result by ______.

   Now, find 5/9 of 729.

4. James bought five computer mice for $36.50.
   Then he sold two of them to his friend.
   How much should he charge his friend?

5. A washer costs $452 and another
   washer costs 3/4 as much.
   Find the price of the other washer.

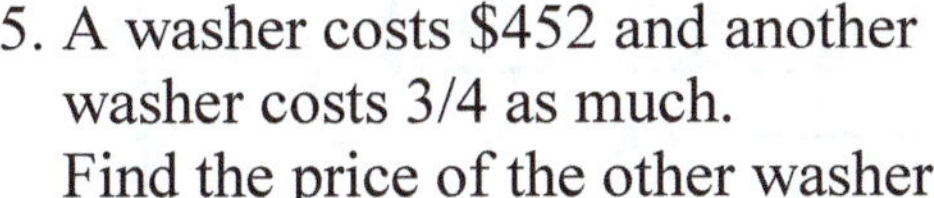

6. The plane had flown $\frac{2}{9}$ of the 12,600-mile trip when the passengers were served supper.

a. How much further was there left to fly?

b. How many miles had they already flown?

7. Mary had $268. She spent $\frac{3}{4}$ of that to buy a bicycle.

a. How much did the bicycle cost?

b. How much does she have left?

8. Twenty tons of flour arrived at the port.
Three tenths of it was sent to New York.
How many tons of flour was that?

How many pounds was it?

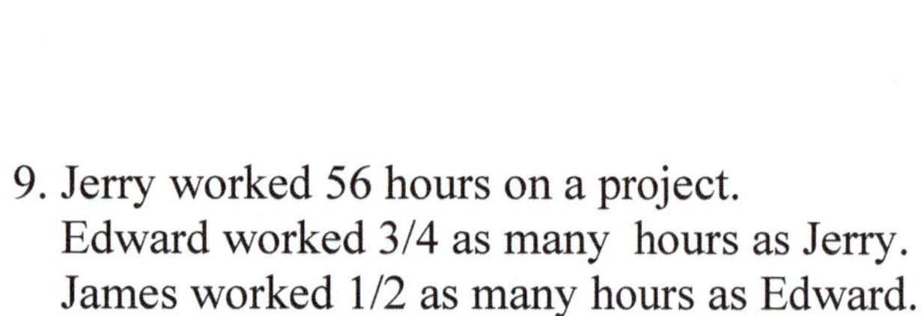

9. Jerry worked 56 hours on a project.
Edward worked 3/4 as many hours as Jerry.
James worked 1/2 as many hours as Edward.

How many hours did the men work altogether?

Write the names and the 56 hours on the diagram. Place the question mark "?" to show what the problem asks to find.

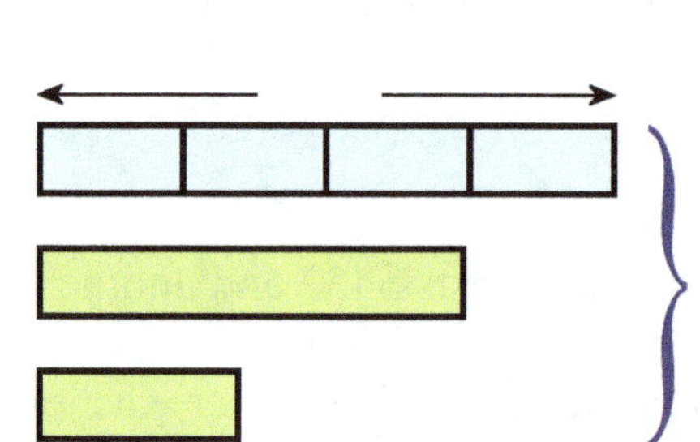

Problems to Solve

Example 1. George has 500 coins in a jar. Some are Canadian coins, some are U.S. coins. There are four times as many U.S. coins as there are Canadian coins. How many U.S. coins does the jar have?

Let's draw a picture to help. If the number of Canadian coins is one "block", then the number of U.S. coins is four "blocks":

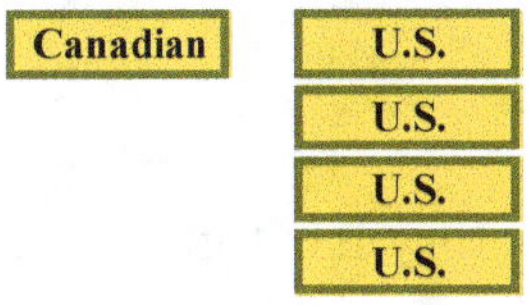

Canadian	U.S.
	U.S.
	U.S.
	U.S.

We now see that the *total* number of coins needs to be divided into *five* equal parts: 500 ÷ 5 = 100.

There are 100 Canadian coins and 400 U.S. coins.

Example 2. Of the 30 children in the bus, 2/5 are girls. How many are boys?

Let's draw a picture to help. Divide the total 30 into *five* parts.

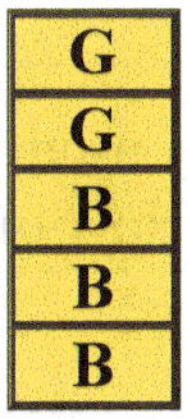

1/5 of 30 is 6 because 30 ÷ 5 = 6.
2/5 of 30 is 12. The rest, or 3/5, are boys. 3/5 of 30 is 18.

There are 18 boys in the bus.

G
G
B
B
B

Example 3. Of the 56 people in the club, 2/7 are children. Of those, 12 are boys. How many are girls?

Divide the total 56 into *seven* parts.

1/7 of 56 is 8. Then, 2/7 of 56 is 16, and that is how many children there are. If 12 of the 16 children are boys, then four are girls.

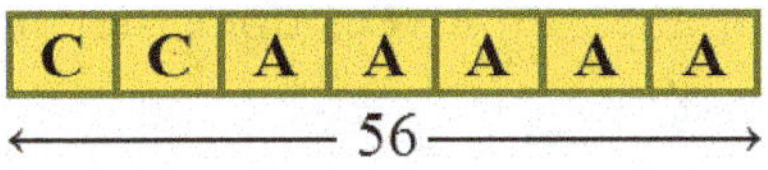

C	C	A	A	A	A	A

←— 56 —→

C = children
A = adults

Example 4. Erica bought a handbag for \$18, which was 2/3 of her money. How much money did she have initially?

In this problem you know the part is \$18 dollars, but you *don't* know the *total*. Divide the total into three parts. Mark the two parts that are \$18.

If two parts are \$18, then one part is \$9, and three parts or the total is \$27. Erica had \$27 to begin with.

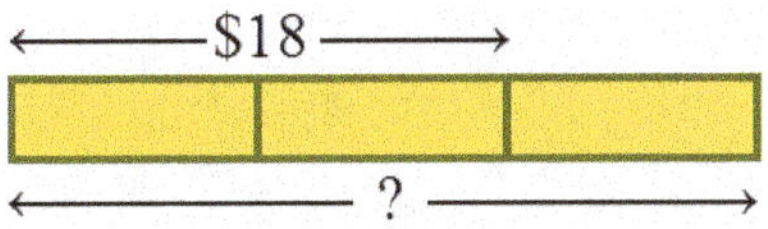

←— \$18 —→

←— ? —→

Solve the problems. You can draw a picture with parts to help.

1. Sally bought some cheap coffee for $3.25 and some specially flavored coffee which was three times as expensive. What was the total cost?

2. A factory has three times as many female workers as male workers. All totaled there are 1,600 workers. How many of them are female?

3. Cindy used half of her allowance to buy a book for $14. How much money does she have left now?

4. One cold day, 1/8 of a company's employees were sick and stayed home; 84 people showed up at work. How many employees does the company have?

5. Mary and Jack shared 24 pieces of candy unequally so that Mary had twice as many pieces as Jack. How many did Mary get?

6. Cindy had 15 yards of ribbon, but she has already used 4/5 of it.

   a. How long is the piece that she has not used?

   b. Of the remaining piece, she cut off 6 inches. How long is the piece that remains now?

7. Of a flock of chickens, 3/4 are white and the rest are speckled. The farmer sold three white chickens and counted that there are still 15 white chickens left. How many speckled chickens are there?

8. Which is a better buy, 1/4 off the price of a CD that costs $12, or 1/3 off the price of a CD for $13.50?

9. Jackie bought 5 notebooks for $1.50 each. They were discounted by one-third. How much did Jackie have to pay?

Divisibility

A number n is **divisible** by another number m, if the division $n \div m$ is exact (no remainder).

For example, 18 ÷ 3 = 6, so 18 is divisible by 3.

Also, 18 is divisible by 6, because we can write the other division 18 ÷ 6 = 3.
So, 18 is divisible by *both* 6 and 3. We say 6 and 3 are **divisors** or **factors** of 18.

You can use long division to check if a number is divisible by another.

For example, 67 ÷ 4 = 16, R3. There is a remainder, so 67 is *not* divisible by 4.

Also, from this we learn that neither 4 nor 16 are divisors of 67.

```
   1 6
4 ) 6 7
  - 4
    2 7
  - 2 4
      3
```

1. Divide and determine if the number is divisible by the other number.

a. 21 ÷ 3 = ______ Is 21 divisible by 3?	**b.** 40 ÷ 6 = ______ Is 40 divisible by 6?	**c.** 17 ÷ 5 = ______ Is 5 a divisor of 17?	**d.** 84 ÷ 7 = ______ Is 7 a factor of 84?

2. Answer the questions. You may need long division.

a. Is 98 divisible by 4?	**b.** Is 603 divisible by 7?	**c.** Is 3 a factor of 1,256?

In any multiplication, the numbers that are multiplied are called **factors** and the result is called a **product**.

factor		factor		product
7	×	6	=	42

For example, since 6 × 7 = 42, 6 and 7 are **factors** of 42.

From this multiplication fact we can write two divisions: 42 ÷ 6 = 7 and 42 ÷ 7 = 6. So, this means that 6 and 7 are also divisors of 42.

From this we can notice the following:

If a number is a factor of another number, it is also its divisor.

There is yet one more new word to learn that ties in with all of this: **multiple**.

We say **42 is a multiple of 6**, because 42 is some number times 6 (namely 7 × 6).

And of course 42 is also a multiple of 7, because 42 is some number times 7 (namely, 6 × 7)!

3. Fill in. We know that 8 × 9 = 72. So, 8 is a ____________________ of 72, and so is 9.

Also, 72 is a ____________________ of 8, and 72 is a ____________________ of 9.

And, 72 is ____________________ by 8 and by 9.

4. Fill in.

a. Is 5 a factor of 55? Yes, because ____ ÷ ____ = ________.	**b.** Is 8 a divisor of 45? No, because ____ ÷ ____ = ___________.
c. Is 36 a multiple of 6? ______, because ____ ÷ ____ = ________.	**d.** Is 34 a multiple of 7? ______, because ____ ÷ ____ = __________.
e. Is 7 a factor of 46? _____, because ______________________.	**f.** Is 63 a multiple of 9? _____, because ____________________.

Multiples of 6 are all those numbers we get when we multiply 6 by other numbers. For example, we can multiply 0 × 6, 7 × 6, 11 × 6, 109 × 6, and so on. The resulting numbers are all multiples of six.

In fact, the skip-counting pattern of 6 gives us a list of multiples of 6:

0, 6, 12, 18, 24, 30, 36, 42, 48, 54, 60, 66, 72, 78, 84, and so on.

5. **a.** Make a list of multiples of 11, starting at 0 and continue at least to 154.

b. Make a list of multiples of 111, starting at 0. Continue as long as you can in this space!

Divisibility by 2

Numbers that are divisible by 2 are called **even** numbers.
Numbers that are NOT divisible by 2 are called **odd** numbers.

Even numbers end in 0, 2, 4, 6, or 8. Every second number is even.

Divisibility by 5

Numbers that end in 0 and 5 are divisible by 5.

For example, 10, 35, 720, and 3,675 are such numbers.

6. Mark an "x" if the number is divisible by 2 or by 5.

number	divisible	
	by 2	by 5
750		
751		
752		
753		
754		

number	divisible	
	by 2	by 5
755		
756		
757		
758		
759		

number	divisible	
	by 2	by 5
760		
761		
762		
763		
764		

number	divisible	
	by 2	by 5
765		
766		
767		
768		
769		

Divisibility by 10

Numbers that end in 0 are divisible by 10.

For example, 10, 60, 340, and 2,570 are such numbers.

7. Mark an "x" if the number is divisible by 2, by 5, or by 10.

number	divisible		
	by 2	by 5	by 10
860			
861			
862			
863			
864			

number	divisible		
	by 2	by 5	by 10
865			
866			
867			
868			
869			

number	divisible		
	by 2	by 5	by 10
870			
871			
872			
873			
874			

If a number is divisible by 10, it ends in a zero, so it is ALSO divisible by ____ and ____.

8. **a.** Write a list of numbers that are divisible by 2, from 0 to 60.

____

This is also a list of __________________________ of 2.

b. In the list above, *underline* those numbers that are divisible by 4.
What do you notice?

c. In the list above, *color* those numbers that are divisible by 6.
What do you notice?

d. Which numbers are divisible by both 4 and by 6?

9. **a.** Write a list of numbers that are divisible by 3, from 0 to 60.

____

This is also a list of __________________________ of 3.

b. In the list above, *underline* those numbers that are divisible by 6.
What do you notice?

c. In the list above, *color* those numbers that are divisible by 9.
What do you notice?

10. Use the lists you made in (8) and (9). Find numbers that are divisible by *both* 2 and 9.

11. What number is a factor of every number?

12. Twenty is a multiple of 4. It is also a multiple of 5. It is also a multiple of four other numbers.
Which ones?

Who am I? (Hint: I am less than 50.) Mystery Number	*Who am I?* (Hint: I am less than 100.) Mystery Number
Divided by 9, I leave a remainder of 6. Divided by 4, I leave a remainder of 1. Divided by 10, I leave a remainder of 3.	I am a multiple of 3, 4, 5, and 6. I am a factor of 120. Divided by 7, I leave a remainder of 4.

Prime Numbers

1. Mark an X if the number is divisible by the given numbers.

number	divisible by 1	divisible by 2	divisible by 3	divisible by 4	divisible by 5	divisible by 6	divisible by 7	divisible by 8	divisible by 9	divisible by 10
2										
3										
4										
5										
6										
7										
8										
9										
10										
11										
12										
13										
14										
15										
16										
17										
18										
19										
20										
21										
22										
23										
24										
25										
26										
27										
28										
29										
30										
31										
32										
33										
34										
35										

2. Now, find each number in this list that is only divisible by 1 and by itself. For example, 7 is one such number: it is only divisible by 1 and by 7. Such numbers are called **primes**.

Prime numbers: ______________________________

A number is prime if the only way to write it as a product is 1 times the number itself.	
For example, 11 is prime, because the only way to write 11 as a product is 1 × 11. But, 12 is not a prime, because we can write it as 2 × 6. We say 12 is **composite**. It is "composed" or "built" from other numbers by multiplication.	
Example 1. Is 450 prime or composite? Since it ends in 0, it is divisible by 10. Indeed, 450 = 45 × 10. So, 450 is composite. **Example 2.** Is 88 prime or composite? Since it is an even number, it is divisible by 2, and we can write it as 88 = 2 × 44. So, 88 is composite.	**Example 3.** Is 37 prime or composite? Check if it is divisible by 2, 3, 4, 5, 6, 7, 8, 9, or 10. It is not divisible by 2. Because of that, it cannot be divisible by 4, 6, 8, or 10 either, so we won't need to check those. It is not divisible by 3 (36 is, and 37 is one more than that). Then it cannot be divisible by 9 either. It is not divisible by 5 since it doesn't end in 0 or 5. It is not divisible by 7. Why? We know 35 is divisible by 7, so 37 leaves a remainder of 2 when divided by 7. So, 37 is prime.

3. Check if each number is prime or composite. If it is composite, write it as a multiplication.

a. 33 is prime/composite If composite: 33 = ____ × ____	**b.** 52 is prime/composite If composite: 52 = ____ × ____	**c.** 41 is prime/composite If composite: 41 = ____ × ____
d. 39 is prime/composite If composite: 39 = ____ × ____	**e.** 43 is prime/composite If composite: 43 = ____ × ____	**f.** 45 is prime/composite If composite: 45 = ____ × ____

Here is one more divisibility rule to help you: **A number is divisible by 3 if the sum of its digits is divisible by 3.**	
Example 4. Is 92 divisible by 3? Add its digits: 9 + 2 = 11. Since 11 is *not* divisible by 3, neither is 92.	**Example 5.** Is 378 divisible by 3? Add its digits: 3 + 7 + 8 = 18. Since 18 *is* divisible by 3, so is 378.

4. Mark an "x" if the number is divisible by 3.

number	digit sum	divisible by 3?
98		
105		
567		
59		

number	digit sum	divisible by 3?
888		
1,045		
1,338		
612		

There is no easy divisibility test for 7. (There is one, but it is not simple to use.) For numbers below 100, you can use the multiplication table of 7, which you should know by heart up to $7 \times 12 = 84$. Beyond that, add 7 to get the two other numbers that are divisible by 7: 91 and 98. You can always use long division to check if a number is divisible by 7.

5. Mark an "x" if the number is divisible by 7.

number	divisible by 7?
99	
74	
56	

number	divisible by 7?
24	
100	
84	

number	divisible by 7?
85	
63	
105	

To check if a number that is between 10 and 100 is prime or composite, it is enough to **check if it is divisible by the primes 2, 3, 5, or 7.** If it is *not* divisible by any of these, it is prime.

6. Are these numbers primes or composites? Use the divisibility rules for 2, 3, and 5 to help you.

a. 67 is prime/composite If composite: 67 = ____ × ____	**b.** 57 is prime/composite If composite: 57 = ____ × ____	**c.** 47 is prime/composite If composite: 47 = ____ × ____
d. 53 is prime/composite If composite: 53 = ____ × ____	**e.** 63 is prime/composite If composite: 63 = ____ × ____	**f.** 61 is prime/composite If composite: 61 = ____ × ____
g. 93 is prime/composite If composite: 93 = ____ × ____	**h.** 85 is prime/composite If composite: 85 = ____ × ____	**i.** 91 is prime/composite If composite: 91 = ____ × ____
j. 87 is prime/composite If composite: 87 = ____ × ____	**k.** 79 is prime/composite If composite: 79 = ____ × ____	**l.** 97 is prime/composite If composite: 97 = ____ × ____

Epilogue: Is 1 a prime number?

Up until 1899, mathematicians listed 1 as a prime number. Since then, modern mathematics has excluded 1 from the list of primes. So in today's books, the list of primes starts from 2. However, even today, some mathematicians insist 1 is a prime.

When 1 is excluded, many theorems and results of mathematics can be written in a simpler way. Fundamentally, the idea of not listing 1 as a prime is a matter of convention and convenience.

Please see also

https://blogs.scientificamerican.com/roots-of-unity/why-isnt-1-a-prime-number/

https://en.wikipedia.org/wiki/Prime_number#Primality_of_one

Finding Factors

Example 1. We can write the number 30 as a multiplication in many different ways:

$30 = 10 \times 3$ and $30 = 2 \times 15$ and $30 = 5 \times 6$. There is yet one more way: $30 = 1 \times 30$.

From this we learn that 10, 3, 2, 15, 5, 6, 1, and 30 are divisors or factors of 30.

What about 7? Well, 30 is *not* divisible by 7, so 7 is not a factor of 30.

It turns out that 1, 2, 3, 5, 6, 10, 15, and 30 are ALL the factors of 30. No other numbers are.

1. Find all the factors of the given numbers. Think of writing the number as a multiplication in many different ways. Don't forget the number itself times 1!

a. 6 factors:	**b.** 10 factors:
c. 12 factors:	**d.** 15 factors:
e. 20 factors:	**f.** 18 factors:

2. These students worked and found all the factors of the given numbers. But is their work correct? Be a teacher detective, and check and correct their work.

a. Aiden found all the factors of 34: $34 = 2 \times 18$ $34 = 1 \times 17$ The factors are 1, 2, 17, 18.	**b.** Olivia found all the factors of 28: $28 = 1 \times 28$ $28 = 2 \times 14$ $28 = 4 \times 7$ The factors are 1, 2, 4, 7, 14, and 28.
c. Jayden found all the factors of 33: $33 = 1 \times 33$ $33 = 3 \times 13$ The factors are 1, 3, 13, 33.	**d.** Isabella found all the factors of 36: $36 = 6 \times 6$ $36 = 4 \times 9$ The factors are 4, 6, and 9.

Example 2. Find all the factors of 85.

Now, it helps to be organized. Let's check if 85 is divisible by all the numbers from 1 to 10.

- It is divisible by 1 (all numbers are): 85 = 1 × 85.
- It is not divisible by 2. Neither by 3 (its digits add up to 13). Of course it can't be divisible by 4, 6, 8, or 10 since it is not even. And it can't be divisible by 9 since it wasn't by 3.
- It *is* divisible by 5. 85 = 5 × 17. And here we can see it is also divisible by 17.
- Is it divisible by 7? No, because 84 is.

Our check is complete. So, we found <u>1, 5, 17, and 85</u>. Those are all the factors of 85.

Why do we <u>not</u> have to check if 85 is divisible by 11, 12, 13, and so on?
Because *if* 85 was 11 times some number, it would be 11 times some *smaller* number than 11. We went through all the smaller numbers already and didn't find that any of them times 11 was 85.

3. Find all the factors of the given numbers.

a. 46 Check 1 2 3 4 5 6 7 8 9 10 factors: ______________________	**b.** 68 Check 1 2 3 4 5 6 7 8 9 10 factors: ______________________
c. 99 Check 1 2 3 4 5 6 7 8 9 10 factors: ______________________	**d.** 72 Check 1 2 3 4 5 6 7 8 9 10 factors: ______________________
e. 73 Check 1 2 3 4 5 6 7 8 9 10 factors: ______________________	**f.** 80 Check 1 2 3 4 5 6 7 8 9 10 factors: ______________________
g. 95 Check 1 2 3 4 5 6 7 8 9 10 factors: ______________________	**h.** 64 Check 1 2 3 4 5 6 7 8 9 10 factors: ______________________

Review

1. Solve.

a.	b.	c.
20 ÷ 10 + 15 = ______	(200 + 100) ÷ 5 = ______	10 × 12 + 40 ÷ 10 = ______
20 × 10 + 15 = ______	200 + 100 ÷ 5 = ______	10 × (12 + 40) ÷ 10 = ______

2. Solve mentally.

a. 3,100 ÷ 100 = _______	**b.** 240 ÷ 20 = _______	**c.** 4,200 ÷ 600 = _______
450 ÷ 10 = _______	800 ÷ 40 = _______	3,200 ÷ 80 = _______

3. Solve.

a.	b.	c.
45 ÷ 6 = ______ R ____	12 ÷ 7 = ______ R ____	31 ÷ 4 = ______ R ____
46 ÷ 6 = ______ R ____	27 ÷ 8 = ______ R ____	56 ÷ 9 = ______ R ____

4. Divide and check your work.

a. 708 ÷ 3 Check:

b. 1,504 ÷ 8 Check:

5. Divide and check your work.

a. 392 ÷ 5 Check:

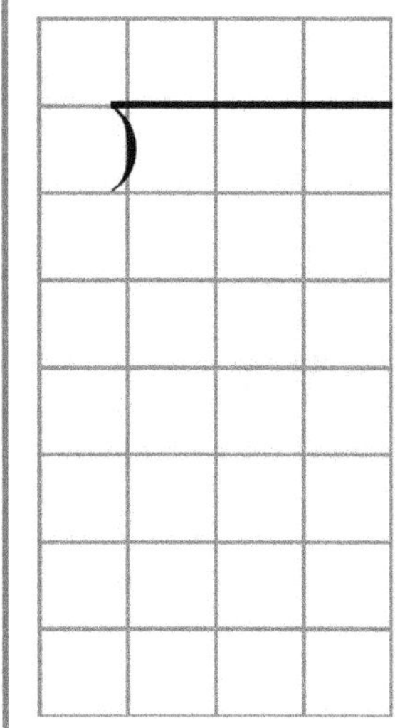

b. 2,845 ÷ 6 Check:

6. Harry has 288 seashells, and Timmy has one fourth of that amount. How many does Timmy have?

7. Mark packaged 70 candles into boxes. Twelve candles fit in each box.

   a. How many boxes were full?

   b. How many candles were in the box that was not full?

8. Four yards of material cost $38.88. How much does one yard cost?

9. John's test scores were 92, 85, 89, 75, and 89. Find his average score.

10. Mark an X if the number is divisible by 3, 5, or 10.

Number	13	40	57	135	354	2,380
Divisible by 3						
Divisible by 5						
Divisible by 10						

11. Fill in.

a. Is 7 a factor of 64? _____, because ____________________________.	**b.** Is 98 a multiple of 2? _____, because ____________________________.
c. Is 76 divisible by 8? _____, because ____________________________.	**d.** Is 30 a factor of 30? _____, because ____________________________.

12. Check if these numbers are primes or composites. Use the divisibility rules for 2, 3, and 5 to help you.

a. 87 is prime/composite If composite: 87 = ____ × ___	**b.** 89 is prime/composite If composite: 89 = ____ × ___	**c.** 91 is prime/composite If composite: 91 = ____ × ___

13. Find all the factors of the given numbers.

a. 24 Check 1 2 3 4 5 6 7 8 9 10 factors: ____________________________	**b.** 27 Check 1 2 3 4 5 6 7 8 9 10 factors: ____________________________
c. 66 Check 1 2 3 4 5 6 7 8 9 10 factors: ____________________________	**d.** 75 Check 1 2 3 4 5 6 7 8 9 10 factors: ____________________________

Puzzle Corner

Imagine you divided all the numbers from 1 to 100 by 6. Which those numbers would have a remainder of 5, when divided by 6?

Math Mammoth Division 2 Answers

Review of Division, p. 7

1. a. $3 \times 4 = 12$; $12 \div 3 = 4$; $12 \div 4 = 3$
   b. $5 \times 3 = 15$; $15 \div 5 = 3$; $15 \div 3 = 5$
   c. $2 \times 4 = 8$; $8 \div 2 = 4$; $8 \div 4 = 2$

2. a. $21 \div 7 = 3$; $21 \div 3 = 7$; $7 \times 3 = 21$; $3 \times 7 = 21$
   b. $24 \div 4 = 6$; $24 \div 6 = 4$; $4 \times 6 = 24$; $6 \times 4 = 24$
   c. $36 \div 4 = 9$; $36 \div 9 = 4$; $9 \times 4 = 36$; $4 \times 9 = 36$

3. a. 8, 9, 10, $22 \div 2 = 11$, $24 \div 2 = 12$, $26 \div 2 = 13$
   b. 9, 8, 7, $30 \div 5 = 6$, $25 \div 5 = 5$, $20 \div 5 = 4$
   c. 9, 10, 11, $120 \div 10 = 12$, $130 \div 10 = 13$, $140 \div 10 = 14$
   d. 8, 7, 6, $35 \div 7 = 5$, $28 \div 7 = 4$, $21 \div 7 = 3$

4.

Eggs	6	12	24	36	42	54	66	78
Omelets	1	2	4	6	7	9	11	13

Thumbtacks	8	24	32	48	64	80	96	104
Pictures	1	3	4	6	8	10	12	13

5. b. $45 – $34 = $11, Jim needs $11 more.
   c. $400 \div 4 = 100$; each box has 100 apples.
   d. $24 \div 6 = 4$; each person got 4 pieces.
   e. $5 \times 50 = 250$ total books.
   f. 2 × $13 = $26; Mom paid $26 for both books.
   g. $20 \div 4 = 5$; there are 5 cows.
   h. $60 \div 3 = 20$; 20 books are on each shelf.

6. a. 9, 10, 5 b. 6, 6, 8 c. 4, 8, 8 d. 8, 3, 5

7. b. $x = 5$ c. $x = 45$ d. $x = 54$

8. a. $10 \times 3 = N$ OR $N = 10 \times 3$; $N = 30$
   b. $9 \times 4 = x$ OR $x = 9 \times 4$; $x = 36$
   c. $20 \times T = 60$, OR $60 = 20 \times T$; $T = 3$
   d. $9 \times y = 81$ OR $81 = y \times 9$; $Y = 9$

9. a. $21 \div 3 = 7$ OR $7 \times 3 = 21$; you can buy 7 books.
   b. $100 \div 5 = 20$ OR $20 \times 5 = 100$; there were 20 apples in each box.
   c. $30 ÷ 5 = $6 OR 5 × $6 = 30; each box costs $6.
   d. $8 \times 5 = 40$; the chocolate bar has 40 squares.
   e. $45 \div 5 = 9$ OR $9 \times 5 = 45$; there are nine fives in 45.
   f. $5 \times 12 = 60$; the boxes weigh 60 pounds.

Division Terms and Division with Zero, p. 10

1. a. 2, the divisor is missing.
   b. 35, the dividend is missing.
   c. 12, the quotient is missing.

2. a. $x \div 7 = 3$; $x = 21$
   b. $140 \div y = 7$; $y = 20$
   c. $150 \div 5 = z$; $z = 30$

3. Answers will vary:
   a. $24 \div 4 = 6$, $30 \div 5 = 6$, $60 \div 10 = 6$
   b. $24 \div 2 = 12$, $24 \div 3 = 8$, $24 \div 6 = 4$

4.

Numbers	Product (written)	Product (solved)	Quotient (written)	Quotient (solved)
12 and 3	12 × 3	36	12 ÷ 3	4
10 and 5	10 × 5	50	10 ÷ 5	2
20 and 4	20 × 4	80	20 ÷ 4	5
100 and 10	100 × 10	1,000	100 ÷ 10	10

5. a. 8, 0, 1
   b. 11, xx, 1
   c. 50, 0, xx
   d. 0, 1, xx

6. a. $x = 64$
   b. $T = 1$
   c. there are many solutions.
   In fact, x can be any number except 0.
   d. $y = 18$

7. Answers will vary. Examples:
   a. $24 \div 24 = 1$, $4 \div 4 = 1$
   b. $0 \div 36 = 0$, $0 \div 12 = 0$

Puzzle corner. The dividend and quotient both were zeros. For example, he could have had the problems $0 \div 6 = 0$ and $0 \div 9 = 0$.

Dividing with Whole Tens and Hundreds, p. 12

1.

a. $300 \times 7 = 2{,}100$ $2100 \div 7 = 300$ $2100 \div 300 = 7$	b. $50 \times 800 = 40{,}000$ $40000 \div 50 = 800$ $40000 \div 800 = 50$	c. $60 \times 40 = 2400$ $2400 \div 60 = 40$ $2400 \div 40 = 60$

2. a. 50, 5, 5, 50
   b. 1,000, 100, 10, 10
   c. 6, 60, 6, 60

3. a. 90, 90
   b. 900, 900
   c. 70, 70

4. a. 40, 4, 400
   b. 9, 90, 90
   c. 60, 60, 6000

5. a. 213 b. 4,022 c. 3,101
   d. 110 e. 1,002 f. 1,410

Finding half...	...is the same as dividing by 2!
$\frac{1}{2}$ of 280 is <u>140</u>	$280 \div 2 =$ <u>140</u>

6. a. 40 b. 12,000 c. 330 d. 2,100

1/2

$806

7. Dad's paycheck was: $806 + $806 = $1,612.

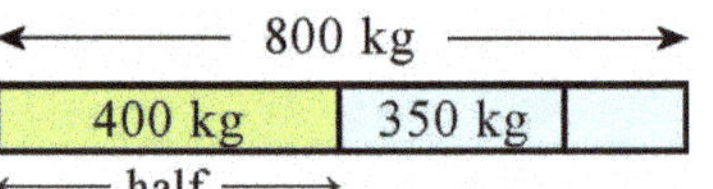

8. The fisherman had: $1/2 \times 800$ kg – 350 kg = 50 kg left.

9. She had $2 \times$ ($12 + $15) = <u>$54</u> in the beginning.

10.

a. $352 \div 5$ $\approx 350 \div 5 = 70$	b. $198 \div 4$ $\approx 200 \div 4 = 50$	c. $403 \div 8$ $\approx 400 \div 8 = 50$

11.

a. $802 \div 21$ $\approx 800 \div 20 = 40$	b. $356 \div 61$ $\approx 360 \div 60 = 6$	c. $596 \div 32$ $\approx 600 \div 30 = 20$

12.

a. $\approx 80 \div 20 = 4$ $\approx 120 \div 60 = 2$ $\approx 2{,}000 \div 500 = 4$	b. $\approx 45 \div 5 = 9$ $\approx 16{,}000 \div 400 = 40$ $\approx 300 \div 30 = 10$

13. $450 \div 5 = 90$

14. Answers will vary but should have a divisor of zero. For example: $67 \div 0$.

15. a. $y = 8{,}000$ b. $s = 4{,}200$ c. $w = 30$

16.

a. $500 \div 5 = 100$ $505 \div 5 = 101$ $510 \div 5 = 102$ $515 \div 5 = 103$ $520 \div 5 = 104$	b. $466 \div 2 = 233$ $468 \div 2 = 234$ $470 \div 2 = 235$ $472 \div 2 = 236$ $474 \div 2 = 237$	c. $366 \div 3 = 122$ $369 \div 3 = 123$ $372 \div 3 = 124$ $375 \div 3 = 125$ $378 \div 3 = 126$

Order of Operations and Division, p. 15

1. a. 3 b. 100 c. 120 d. 2,000

2. a. 62 b. 152 c. 2,000 d. 18

3. a. 9 b. 17 c. 200 d. 5

4.

a.	b.	c.
$24 \div 2 + 10 = 22$ $24 \div (2 + 10) = 2$	$18 + 30 \div 2 = 33$ $(18 + 30) \div 2 = 24$	$40 - 40 \div 8 = 35$ $(40 - 40) \div 8 = 0$

5. a. $(20 + 15) \div 5 = 7$
   b. $20 - 50 \div 5 = 10$
   c. $20 \times 30 - 100 = 500$

6. $(21 + 17) \div 2$. The answer is 19 figures.

7. $6 \times 6 \div 4$. The answer is \$9.

8. a. 5; 7 b. 60; 120 c. 20; 20 d. 1; 1 e. 0; 0

9. a. $5 \div 5 \times 5 = 5$
   b. $(5 - 5) \times 5 = 0$
   c. $(5 + 5) \div 5 = 2$
   d. $(5 + 5) \times (5 + 5) = 100$
   e. $5 \times 5 + 5 - 5 = 25$
   OR $5 \times 5 - 5 + 5 = 25$
   OR $5 - 5 + 5 \times 5 = 25$
   OR $5 \times 5 - (5 - 5) = 25$

Puzzle corner:
$(5 - 5) \times 5 = 0$
$5 \div 5 = 1$
$(5 + 5) \div 5 = 2$
$(5 + 5 + 5) \div 5 = 3$
$(5 \times 5 - 5) \div 5 = 4$
$5 \times 5 \div 5 = 5$
$(5 \times 5 + 5) \div 5 = 6$
$(5 \times 5 + 5 + 5) \div 5 = 7$
$(5 + 5) - (5 + 5) \div 5 = 8$
$(5 + 5) - (5 \div 5) = 9$
$5 + 5 = 10$

The Remainder, Part 1, p. 17

1.

a. Divide into groups of 4.	b. Divide into groups of 2.	c. Divide into groups of 5.
	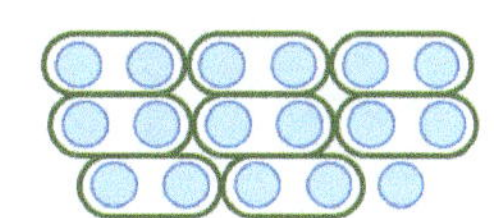	
$10 \div 4 = 2$ R2	$17 \div 2 = 8$ R1	$12 \div 5 = 2$ R2

2. a. $14 \div 4 = 3$ R2 b. $7 \div 3 = 2$ R1 c. $19 \div 6 = 3$ R1

3.

a. Divide 16 into groups of 5.	b. Divide 17 into groups of 3.	c. Divide 15 into groups of 4.

$16 \div 5 = 3$ R1	$17 \div 3 = 5$ R2	$15 \div 4 = 3$ R3

4.

a. $17 \div 4 = 4$ R1	b. $9 \div 2 = 4$ R1	c. $11 \div 6 = 1$ R5

5. a. $10 \div 3 = 3$ R1 b. $17 \div 5 = 3$ R2 c. $11 \div 4 = 2$ R3

6. a. $13 \div 5 = 2$ R3 b. $18 \div 4 = 4$ R2 c. $10 \div 4 = 2$ R2

The Remainder, Part 1, continued

7.

a. 27 ÷ 5 = 5 R2 5 goes into 27 five times.	b. 16 ÷ 6 = 2 R4 6 goes into 16 two times.	c. 11 ÷ 2 = 5 R1 2 goes into 11 five times.
d. 37 ÷ 5 = 7 R2	e. 26 ÷ 3 = 8 R2	f. 56 ÷ 9 = 6 R2
g. 43 ÷ 5 = 8 R3	h. 34 ÷ 6 = 5 R4	i. 40 ÷ 7 = 5 R5

8.

a.	b.	c.
23 ÷ 4 = 5 R3 23 ÷ 5 = 4 R3	16 ÷ 7 = 2 R2 20 ÷ 3 = 6 R2	21 ÷ 8 = 2 R5 12 ÷ 9 = 1 R3

9.

a. 10 ÷ 5 = 2 R0 11 ÷ 5 = 2 R1 12 ÷ 5 = 2 R2 13 ÷ 5 = 2 R3 14 ÷ 5 = 2 R4 15 ÷ 5 = 3 R0	b. 17 ÷ 3 = 5 R2 18 ÷ 3 = 6 R0 19 ÷ 3 = 6 R1 20 ÷ 3 = 6 R2 21 ÷ 3 = 7 R0 22 ÷ 3 = 7 R1	c. 12 ÷ 4 = 3 R0 13 ÷ 4 = 3 R1 14 ÷ 4 = 3 R2 15 ÷ 4 = 3 R3 16 ÷ 4 = 4 R0 17 ÷ 4 = 4 R1

10. a. 27 ÷ 5 = 5 R2. He had five rows. Two cars were left over.
b. 19 ÷ 5 = 3 R4. She had five groups of 5. You can make a smaller group with only four children in it.
c. (36 − 3) ÷ 6 = 5 R3. She has five full bags of cookies.
d. No, because 51 ÷ 8 = 6 R3.
e. Of four, no. 35 ÷ 4 = 8 R3 (the division is not even.) Of five, yes. 35 ÷ 5 = 7.
Of six, no. 35 ÷ 6 = 5 R5. Of seven, yes. 35 ÷ 7 = 5.
f. 38 ÷ 6 = 6 R2. There were two photos on the last page. Six pages were full.

The Remainder, Part 2, p. 20

1. a. 3 b. 9 c. 7 d. 9

2. a. $\begin{array}{r} 6 \\ 5\overline{)32} \\ -30 \\ \hline 2 \end{array}$ b. $\begin{array}{r} 8 \\ 5\overline{)44} \\ -40 \\ \hline 4 \end{array}$ c. $\begin{array}{r} 6 \\ 6\overline{)37} \\ -36 \\ \hline 1 \end{array}$ d. $\begin{array}{r} 4 \\ 7\overline{)29} \\ -28 \\ \hline 1 \end{array}$

e. $\begin{array}{r} 5 \\ 8\overline{)46} \\ -40 \\ \hline 6 \end{array}$ f. $\begin{array}{r} 5 \\ 9\overline{)52} \\ -45 \\ \hline 7 \end{array}$ g. $\begin{array}{r} 8 \\ 4\overline{)35} \\ -32 \\ \hline 3 \end{array}$ h. $\begin{array}{r} 6 \\ 9\overline{)57} \\ -54 \\ \hline 3 \end{array}$

3. a. 6 × 5 + 2 = 32 b. 8 × 5 + 4 = 44 c. 6 × 6 + 1 = 37 d. 4 × 7 + 1 = 29
e. 5 × 8 + 6 = 46 f. 5 × 9 + 7 = 52 g. 8 × 4 + 3 = 35 h. 6 × 9 + 3 = 57

4. 33 ÷ 6 = 5 R3. Jill needed six containers, but only five were full.

5. 53 ÷ 12 = 4 R5. Mom needed 5 cartons for all the eggs.

6. 36 ÷ 11 = 3 R3. She put three pencils back into the cabinet.

7. 58 ÷ 8 = 7 R2. They got seven full boxes.

8. 3 × 23 + 15 = 84. He had 84 award stickers.

The Remainder, Part 3, p. 22

1. One bus holds 42 children. Two buses hold 84 children. Three buses hold 126 children. So, three buses were needed to hold 100 children

2. She needed four folders. (Three folders is not enough, because $3 \times 20 = 60$. Yet, four *is* enough because $4 \times 20 = 80$.) Three of them were full.

3. a. They could make three classes of 22 first graders. (They don't have enough first graders for four classes since $4 \times 22 = 88$ which is more than 77.)

   b. They will get three classes of 20 first graders, and 17 children in a fourth class.

4. The teams had six, seven, and seven players.

5. a. 7 R5 b. 8 R2 c. 8 R4 d. 4 R3

6.

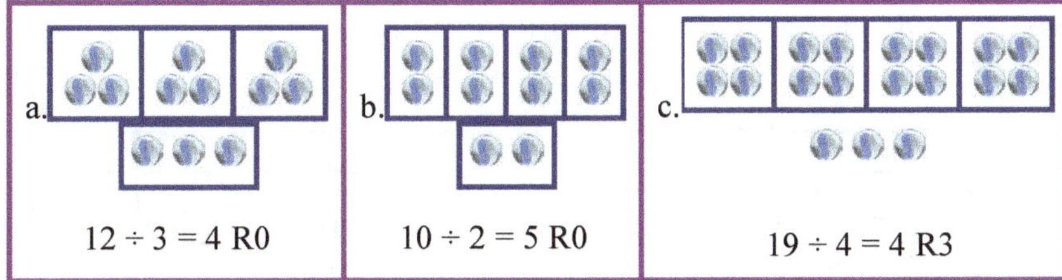

7.

a.	b.	c.
a. $21 \div 5 = 4$ R1 $22 \div 5 = 4$ R2 $23 \div 5 = 4$ R3 $24 \div 5 = 4$ R4	b. $56 \div 8 = 7$ R0 $57 \div 8 = 7$ R1 $58 \div 8 = 7$ R2 $59 \div 8 = 7$ R3	c. $43 \div 7 = 6$ R1 $44 \div 7 = 6$ R2 $45 \div 7 = 6$ R3 $46 \div 7 = 6$ R4

8. The shortcut is: the remainder is always the last digit of the dividend (the number you divide), and the other digits are the quotient (the answer).

a.	b.	c.
a. $29 \div 10 = 2$ R9 $30 \div 10 = 3$ R0 $31 \div 10 = 3$ R1	b. $78 \div 10 = 7$ R8 $79 \div 10 = 7$ R9 $80 \div 10 = 8$ R0	c. $54 \div 10 = 5$ R4 $55 \div 10 = 5$ R5 $56 \div 10 = 5$ R6

Puzzle Corner:
a. $16 \div 5 = 3$ R1 OR $16 \div 3 = 5$ R1
b. $31 \div 7 = 4$ R3 OR $31 \div 4 = 7$ R3
c. $135 \div 30 = 4$ R3 OR $123 \div 4 = 30$ R3

Long Division 1, p. 24

1.

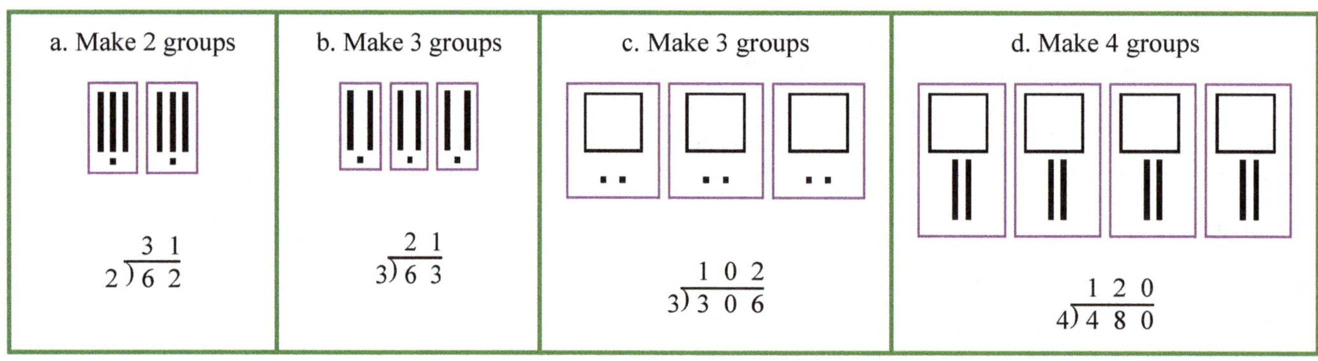

2. a. 21 b. 131 c. 220 d. 2,010 e. 22 f. 3,021 g. 110 h. 1,201

3. a. 41 b. 71 c. 60 d. 31 e. 92 f. 61 g. 611 h. 601 i. 710 j. 901

4.

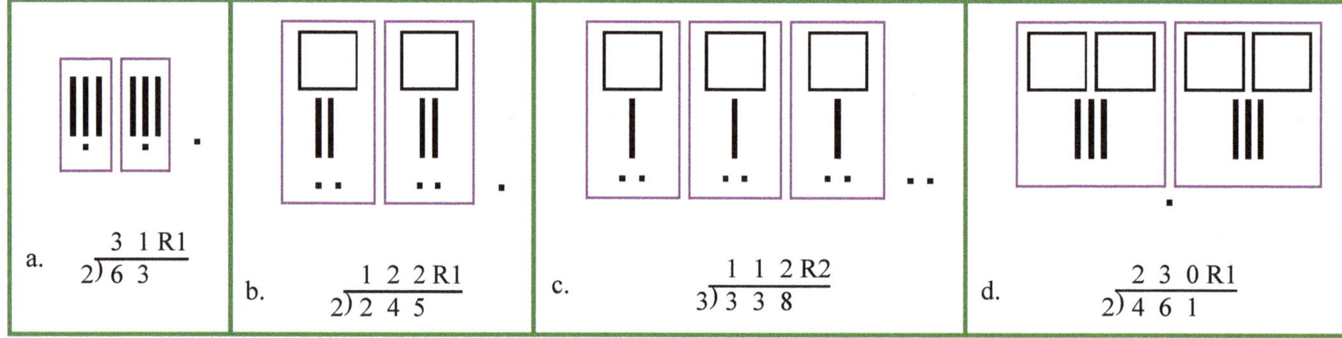

5. a. 211 R3 b. 34 R1 c. 122 R1 d. 22 R1 e. 60 R1 f. 300 R5 g. 30 R5 h. 310 R2

6. a. 42 R2 b. 31 R2 c. 711 R1 d. 711 R1 e. 1,101 R2 f. 4,031 R1

7. a. 110, 410 b. 9, 123 c. 412, 6 R20

Long Division 2, p. 28

1. a. 16 b. 24 c. 29 d. 15 e. 19 f. 39 g. 26 h. 47

2. a. 47 b. 43 c. 64 d. 34 e. 84 f. 58

Long Division 3, p. 31

1. a. 115 b. 123 c. 244 d. 276 e. 318 f. 121
 g. 113 h. 113 i. 325 j. 113 k. 112 l. 218

2. a. 189 b. 166 c. 142 d. 117 e. 152 f. 117

Long Division with 4-Digit Numbers, p. 35

1. a. 2,347 b. 2,310 c. 1,785 d. 4,885

2. a. 1,934 b. 551 c. 1,340
 d. 1,138 e. 1,317 f. 1,216

3. a. 493 b. 384 c. 924 d. 49 e. 87 f. 371

4. 9 × \$16 ÷ 2 = \$72. They each paid \$72.

5. 504 min ÷ 7 = 72 min, or 1 h 12 min each day

6. a. 2600 ÷ 8 = 325. The second clue is at 325 feet.
 b. The third clue is at 650 feet.

7. a. 96 ÷ 6 = 16. Sixteen children were coming to the party.
 b. 8 × 25 = 200 and 200 − 96 = 104.
 She had 104 balloons left.

More Long Division, p. 39

1. a. 1,045 b. 1,406 c. 2,037 d. 1,307
2. a. 2,705 b. 1,308 c. 1,309 d. 1,063
3. a. 108 b. 205 c. 402 d. 405 e. 308 f. 1,070
4. a. $285 \div 5 = 57$.
   There are 57 buttons in one compartment.
   b. $3 \times 57 = 171$.
   There are 171 buttons in three compartments.
   c. $4 \times 57 = 228$.
   There are 228 buttons in four compartments.
5. The payments were ($9,620 − $2,000) ÷ 4 = $1,905 each.
6. a. 21,234 b. 35,407 c. 21,645 d. 3,162 e. 5,275

Remainder Problems, p. 42

1. a. 171 R1 Check: $3 \times 171 + 1 = 514$
   b. 84 R1 Check: $8 \times 84 + 1 = 673$
   c. 317 R3 Check: $6 \times 317 + 3 = 1{,}905$
   d. 2,051 R1 Check: $4 \times 2{,}051 + 1 = 8{,}205$
2. a. wrong; 77 R1 b. right c. wrong: 451
   d. The remainder is larger than the divisor.
3. a. $112 \div 9 = 12$ R4. We get 12 rows, 9 chairs each row, and 4 chairs will be left over or put in an extra row.
   b. $800 \div 3 = 266$ R2. We get 266 piles, 3 erasers in each pile, and 2 erasers left over.
4. They will get 166 full bags.
5. $20 \times 50 = 1{,}000$ and $19 \times 50 = 950$. So, 19 buses is enough to transport 940 people.
6. $75 \div 4 = 18$ R3. One 18-day vacation and three 19-day vacations. If the division had been even, all of the vacations would have been 18 days, but now there are three extra days to be added to three of the vacations.
7. $400 - (2 \times 90) - (4 \times 40) = 60$; $60 \div 6 = 10$.
   They will have ten full 6-kg boxes of strawberries.
8. a. Yes. There will be 103 containers. $412 \div 4 = 103$.
   b. No, there will be 82 containers with 2 left over.
   $412 \div 5 = 82$ R2.
   c. No, there will be 68 with four left over.
   $412 \div 6 = 68$ R4.
9. $740 \div 6 = 123$ R2. Paint 123 pieces in four of the colors (any four), and 124 pieces in the two remaining colors.
10. a. 70 R1; 70 R2; 71
    b. 172 R2; 172 R3; 172 R4
    c. 82 R1; 82 R2; 82 R3
    d. 798 R3 798 R4 798 R5
    You can figure out the two other problems after solving one, because the remainder will increase by one as the dividend increases by one.
11. It would be 38 R4. The only difference is that the remainder increases by 1.
12. a. 78 R7; 6 R6; 34
    b. 45 R2; 50 R9; 5 R2
    c. 46 R3; 98 R2; 92 R5
    The ones digit of the dividend will always be the remainder.

Puzzle Corner: The remainder is larger than the divisor.

Long Division with Money, p. 46

1. a. $8.47 b. $3.72
2. a. $28.50 b. $1.14
3. $25.56 + $3.55 + $2.75 = $31.86
   $31.86 ÷ 2 = $15.93. Each girl paid $15.93.
4. ($25.95 + $4.35) ÷ 3 = $10.10. Each person's share was $10.10.
5. $358.60 − $100 = $258.60; $258.60 ÷ 4 = $64.65.
   Each payment was $64.65.
6. $12.96 ÷ 8 = $1.62.
   You will pay $1.62. Your brother will pay $1.62.
   Mom will pay $12.96 − $1.62 − $1.62 = $9.72.

Long Division Crossword Puzzle, p. 48

1. Across:
   a. 3,440 ÷ 8 = 430
   b. 574 ÷ 7 = 82
   c. 234 ÷ 9 = 26
   d. 1,707 ÷ 3 = 569
   e. 4,756 ÷ 2 = 2,378

   Down:
   a. 1,072 ÷ 8 = 134
   b. 6,135 ÷ 3 = 2,045
   c. 145 ÷ 5 = 29
   d. 2,652 ÷ 4 = 663
   e. 1,442 ÷ 7 = 206
   f. 3,474 ÷ 9 = 386

a. 1					
3		**b.** 2		**b.** 8	**e.** 2
a. 4	3	0			0
		4		**c.** 2	6
		d. 5	**d.** 6	9	
			6		**f.** 3
		e. 2	3	7	8
					6

Average, p. 49

1. (78 + 87 + 69 + 86) ÷ 4 = 80. Judith's average score is 80.

2. (18 + 22 + 26 + 23 + 16) ÷ 5 = 21. The average temperature for the day was 21°C.

3. 414 ÷ 6 = 69. Dad averaged 69 km in one hour.

4. 12 × 55 = 660. A dozen eggs would weigh 660 grams.

5. 7 × 76 = 532. It cost $532 for one week.

6. (234 + 178 + 250 + 198) ÷ 4 = 215. Her weekly average grocery bill was $215.

7. The girls' average time was 15 minutes. The boys' average time was 13 minutes. The boys are faster. The difference is two minutes.

8. a.

Quiz score	Frequency
13-15	1
16-18	1
19-21	2
22-24	4
25-27	0
28-30	2

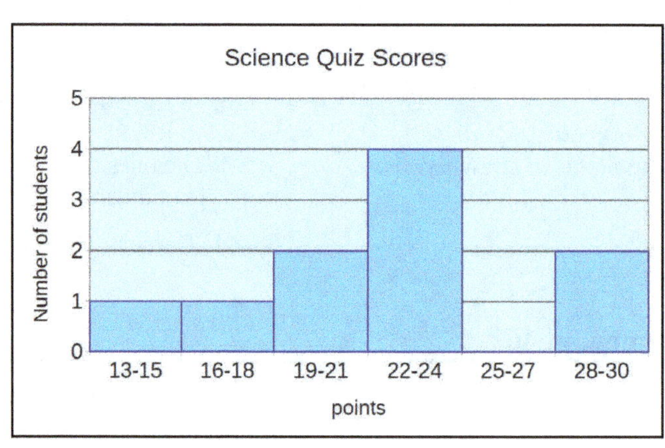

   b. The average score is 22.
   c. Look at the "peak" of the graph. The average is usually near that point.

9. a. The average age is 29. b. Now the average age is 34.

Puzzle corner: 213 ÷ 12 is 17 R9.

Finding Fractional Parts with Division, p. 52

1.

a. 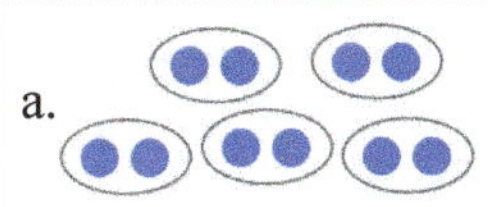	b.	c.	d.
$10 \div 5 = 2$	$9 \div 3 = 3$	$16 \div 2 = 8$	$15 \div 3 = 5$
$\frac{1}{5}$ of 10 is 2.	$\frac{1}{3}$ of 9 is 3.	$\frac{1}{2}$ of 16 is 8.	$\frac{1}{3}$ of 15 is 5.

2.

a. $30 \div 5 = 6$	b. $48 \div 6 = 8$	c. $25 \div 5 = 5$	d. $50 \div 5 = 10$
$\frac{1}{5}$ of 30 is 6.	$\frac{1}{6}$ of 48 is 8.	$\frac{1}{5}$ of 25 is 5.	$\frac{1}{5}$ of 50 is 10.

3.

a. $\frac{1}{6}$ of 30 is 5.	b. $\frac{1}{7}$ of 49 is 7.	c. $\frac{1}{10}$ of 250 is 25.
$30 \div 6 = 5$	$49 \div 7 = 7$	$250 \div 10 = 25$
d. $\frac{1}{2}$ of 480 is 240.	e. $\frac{1}{9}$ of 1,800 is 200.	f. $\frac{1}{5}$ of 400 is 80.
$480 \div 2 = 240$	$1{,}800 \div 9 = 200$	$400 \div 5 = 80$

4.

a.	b.	c.
$\frac{1}{3}$ of 9 apples is 3 apples.	$\frac{1}{4}$ of 12 flowers is 3.	$\frac{1}{5}$ of 15 fish is 3 fish.
$\frac{2}{3}$ of 9 apples is 6 apples.	$\frac{2}{4}$ of 12 flowers is 6.	$\frac{2}{5}$ of 15 fish is 6 fish.
$\frac{3}{3}$ of 9 apples is 9 apples.	$\frac{3}{4}$ of 12 flowers is 9.	$\frac{3}{5}$ of 15 fish is 9 fish.
	$\frac{4}{4}$ of 12 flowers is 12.	$\frac{4}{5}$ of 15 fish is 12 fish.
		$\frac{5}{5}$ of 15 fish is 15 fish.

5. a. 4, 8, 12   b. 4, 12, 20   c. 50, 150, 350
d. 70, 140, 350   e. 30, 210, 330   f. 7, 63, 350

6. a. Marsha got \$18 from her mom. She put into her savings \$6, which was one-third part of it. $18 \div 6 = 3$
b. Mariana spent one-fourth of her \$80 savings, or \$20. $80 \div 4 = 20$
c. One-fifth of all the 25 boys went jogging, which meant that 5 boys went jogging. $25 \div 5 = 5$

7. a. One pound is one-eighth part of the bag. It costs \$0.40.
b. Five-eighths costs \$2.00.

8. a. One-tenth of the pie weighs 120 grams.
b. Nine-tenths of the pie weighs 1,080 grams.

9. $28 \div 4 = 7$, and $7 \times 3 = 21$. Three would cost \$21.

Finding Fractional Parts with Division, cont.

10. You can solve this in several ways. One way is to use the idea that ¼ is half of ½ and that ⅛ is half of ¼. The other way is to calculate ½ of $24.40 separately, ¼ of $24.40 separately, and 1/8 of $24.40 separately.

Mark: $12.20 Judy: $6.10 Art: $3.05 Grace: $3.05

11. Erica and James each had 28 balloons to sell. By the evening, Erica had sold 1/2 or 14 of her balloons. James had sold 3/4 or 21 of his. Together they had sold 35 balloons.

Problems with Fractional Parts, p. 55

1. a. One slice weighs 20 grams.
   b. Three slices weigh 60 grams.
   c. Eleven slices weigh 220 grams.

2. a. Two-sixths of the pie weighs 150 grams.
   b. It weighs 375 grams.

3. If you need to calculate 5/9 of the number 729, first divide 729 by 9, then multiply the result by 5. 5/9 of 729 is 405

4. $36.50 ÷ 5 × 2 = $14.60

5. The other washer costs $452 ÷ 4 × 3 = $339.

6. 12,600 ÷ 9 × 2 = 2,800
   a. 9,800 miles left b. 2,800 miles

7. $268 ÷ 4 × 3 = $201. It cost $201. She has $67 left.

8. a. 6 tons, or 12,000 pounds.

9. Edward worked (56 ÷ 4) × 3 = 42 hours.
   James worked 21 hours.
   Together they worked 56 + 42 + 21 = 119 hours.

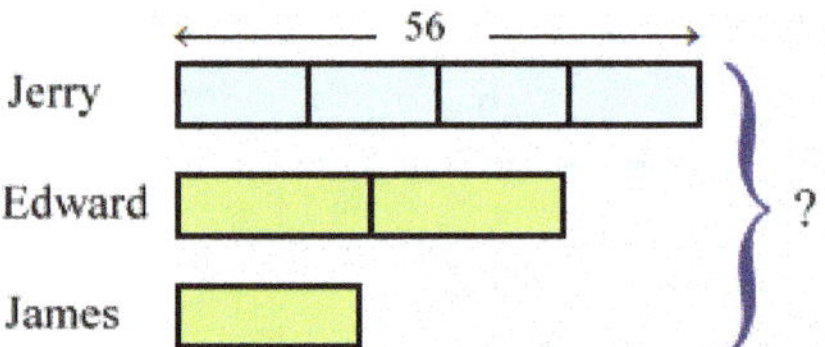

Problems to Solve, p. 57

1. $3.25 + 3 × $3.25 = $13

2. 1,200 are females. Since there are three times as many females as males, we can divide the 1,600 workers into four parts. One-fourth part of 1,600 is 400. So, there are 3 × 400 or 1,200 female workers.

3. Cindy has $14 left. (Half of Cindy's money is $14.)

4. 96 workers. Since 84 is 7/8 of the workers, 84 ÷ 7 = 12 gives you 1/8 of the workers, and then 8 × 12 = 96 is the total amount.

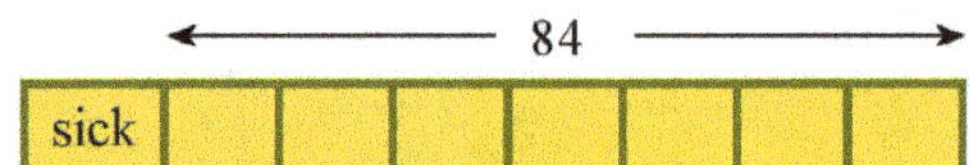

5. Mary got 16 pieces. One-third of the pieces is 8 pieces.

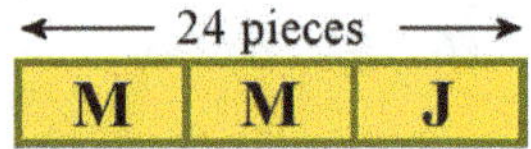

6. 2 yards, 2 feet, 6 inches.
   One-fifth of 15 yards is 3 yards. Now subtract 3 yards − 6 inches = 2 yards 2 feet 6 inches.

7. There are 6 speckled chickens. Solution: there are 18 white chickens (three that were sold and 15 that were left), and that is 3/4 of all the chickens.

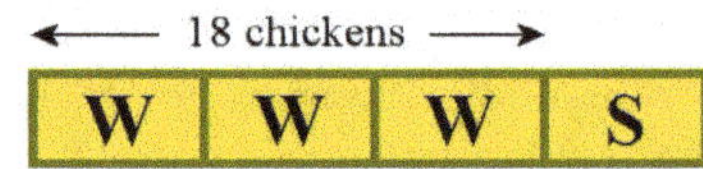

   So 18 ÷ 3 = 6 gives you 1/4 of the chickens (one block). And then, one block, or six chickens, are speckled.

8. They would both cost $9 after the discount.
   One-fourth of $12 is $3, so the new price is $9.
   One-third of $13.50 is $4.50, so the new price is $9.

9. Jackie paid $5.00. First find the total without the discount: 5 × $1.50 = $7.50. One-third of that is $2.50. The price with discount is then $7.50 − $2.50 = $5.00

Divisibility, p. 60

1. a. 7; yes b. 6 R4; no c. 3 R2; no d. 12; yes

2. a. 24 R2, no b. 86 R1; no c. 418 R2; no

3. Here is a multiplication fact: 8 × 9 = 72. So, 8 is a factor of 72, and so is 9.
   Also, 72 is a multiple of 8, and also 72 is a multiple of 9. And, 72 is divisible by 8 and also by 9.

Divisibility, continued

4.

a. Is 5 a factor of 55? Yes, because 5 × 11 = 55.	b. Is 8 a divisor of 45? No, because 45 ÷ 8 = 5 R5.
c. Is 36 a multiple of 6? Yes, because 6 × 6 = 36.	d. Is 34 a multiple of 7? No, because 34 ÷ 7 = 4 R6.
e. Is 7 a factor of 46? No, because 46 ÷ 7 = 6 R4. (It is not an even division.)	f. Is 63 a multiple of 9? Yes, because 7 × 9 = 63.

5. a. 0, 11, 22, 33, 44, 55, 66, 77, 88, 99, 110, 121, 132, 143, 154
   b. 0, 111, 222, 333, 444, 555, 666, 777, 888, 999, 1,110, 1,221, 1,332, 1,443, 1,554, 1,665

6.

number	divisible	
	by 2	by 5
750	x	x
751		
752	x	
753		
754	x	

number	divisible	
	by 2	by 5
755		x
756	x	
757		
758	x	
759		

number	divisible	
	by 2	by 5
760	x	x
761		
762	x	
763		
764	x	

number	divisible	
	by 2	by 5
765		x
766	x	
767		
768	x	
769		

7.

number	divisible		
	by 2	by 5	by 10
860	x	x	x
861			
862	x		
863			
864	x		

number	divisible		
	by 2	by 5	by 10
865		x	
866	x		
867			
868	x		
869			

number	divisible		
	by 2	by 5	by 10
870	x	x	x
871			
872	x		
873			
874	x		

If a number is divisible by 10, it ends in zero, so it is ALSO divisible by 2 and 5.

8. a. 2, 4, 6, 8, 10, 12, 14, 16, 18, 20, 22, 24, 26, 28, 30, 32, 34, 36, 38, 40, 42, 44, 46, 48, 50, 52, 54, 56, 58, 60
   This is also a list of multiples of (or multiplication table of) 2.

   b. 2, 4, 6, 8, 10, 12, 14, 16, 18, 20, 22, 24, 26, 28, 30, 32, 34, 36, 38, 40, 42, 44, 46, 48, 50, 52, 54, 56, 58, 60
   These are every other number in the list of multiples of 2.

   c. 2, 4, **6**, 8, 10, **12**, 14, 16, **18**, 20, 22, **24**, 26, 28, **30**, 32, 34, **36**, 38, 40, **42**, 44, 46, **48**, 50, 52, **54**, 56, 58, **60**
   These are every third number in the list of multiples of 2, or every third even number divisible by 6.

   d. 12, 24, 36, 48, and 60 - or multiples of 12.

9. a. 3, 6, 9, 12, 15, 18, 21, 24, 27, 30, 33, 36, 39, 42, 45, 48, 51, 54, 57, 60
   This is also a list of multiples of (or multiplication table of) 3.

   b. 3, 6, 9, 12, 15, 18, 21, 24, 27, 30, 33, 36, 39, 42, 45, 48, 51, 54, 57, 60
   These are every second number in the list of multiples of 3.

   c. 3, 6, **9**, 12, 15, **18**, 21, 24, **27**, 30, 33, **36**, 39, 42, **45**, 48, 51, **54**, 57, 60
   These are every third number in the list of multiples of 3.

10. 18, 36, 54

11. 1

12. It is also a multiple of 1, 2, 10, and 20.

Mystery number: 33 and 60

Prime Numbers, p. 64

number	divisible by 1	divisible by 2	divisible by 3	divisible by 4	divisible by 5	divisible by 6	divisible by 7	divisible by 8	divisible by 9	divisible by 10
2	x	x								
3	x		x							
4	x	x		x						
5	x				x					
6	x	x	x			x				
7	x						x			
8	x	x		x				x		
9	x		x						x	
10	x	x			x					x
11	x									
12	x	x	x	x		x				
13	x									
14	x	x					x			
15	x		x		x					
16	x	x		x				x		
17	x									
18	x	x	x			x			x	
19	x									
20	x	x		x	x					x
21	x		x				x			
22	x	x								
23	x									
24	x	x	x	x		x		x		
25	x				x					
26	x	x								
27	x		x						x	
28	x	x		x			x			
29	x									
30	x	x	x		x	x				x
31	x									
32	x	x		x				x		
33	x		x							
34	x	x								
35	x				x		x			

2. Prime numbers: 2, 3, 5, 7, 11, 13, 17, 19, 23, 29, 31

3. Answers will vary, as you can write a composite number as a product in many different ways.

a. 33 is composite. $33 = 3 \times 11$	b. 52 is composite. $52 = 2 \times 26$	c. 41 is prime.
d. 39 is composite. $39 = 3 \times 13$	e. 43 is prime.	f. 45 is composite. $45 = 5 \times 9$

Prime Numbers, continued

4.

number	digit sum	divisible by 3?
98	17	no
105	6	yes
567	18	yes
59	14	no

number	digit sum	divisible by 3?
888	24	yes
1,045	10	no
1,338	15	yes
612	9	yes

5.

number	divisible by 7?
99	no
74	no
56	yes

number	divisible by 7?
24	no
100	no
84	yes

number	divisible by 7?
85	no
63	yes
105	yes

6. Answers will vary, as you can write a composite number as a product in many different ways.

a. 67 is prime.	b. 57 is composite. 57 = 3 × 19	c. 47 is prime.
d. 53 is prime.	e. 63 is composite. 63 = 7 × 9	f. 61 is prime.
g. 93 is composite. 93 = 3 × 31	h. 85 is composite. 85 = 5 × 17	i. 91 is composite. 91 = 7 × 13
j. 87 is composite. 87 = 3 × 29	k. 79 is prime.	l. 97 is prime.

Finding Factors, p. 67

1.

a. factors: 1, 2, 3, 6	b. factors: 1, 2, 5, 10
c. factors: 1, 2, 3, 4, 6, 12	d. factors: 1, 3, 5, 15
e. factors: 1, 2, 4, 5, 10, 20	f. factors: 1, 2, 3, 6, 9, 18

2. Only Olivia's work was totally correct.

a. Aiden found all the factors of 34: ~~34 = 2 × 18~~ 34 = 2 × 17 ~~34 = 1 × 17~~ 34 = 1 × 34 The factors are 1, 2, 17, ~~18~~, 34	b. Olivia found all the factors of 28: 28 = 1 × 28 28 = 2 × 14 28 = 4 × 7 The factors are 1, 2, 4, 7, 14, and 28.
c. Jayden found all the factors of 33: 33 = 1 × 33 ~~33 = 3 × 13~~ 33 = 3 × 11 The factors are 1, 3, ~~13~~, 11, 33.	d. Isabella found all the factors of 36: 36 = 6 × 6 36 = 3 × 12 36 = 3 × 12 36 = 4 × 9 36 = 1 × 36 The factors are 4, 6, and 9. Also 1, 2, 3, 12, 18, 36

Finding Factors, continued

3.

a. factors: 1, 2, 23, 46	b. factors: 1, 2, 4, 17, 34, 68
c. factors: 1, 3, 9, 11, 33, 99	d. factors: 1, 2, 3, 4, 6, 8, 9, 12, 18, 24, 36, 72
e. factors: 1, 73	f. factors: 1, 2, 4, 5, 8, 10, 16, 20, 40, 80
g. factors: 1, 5, 19, 95	h. factors: 1, 2, 4, 8, 16, 32, 64

Mixed Review Chapter 5, p. 69

1. a. 4,284   b. 49,068

2. a. 84; 80   b. 20; 54   c. 1,090; 90

3.

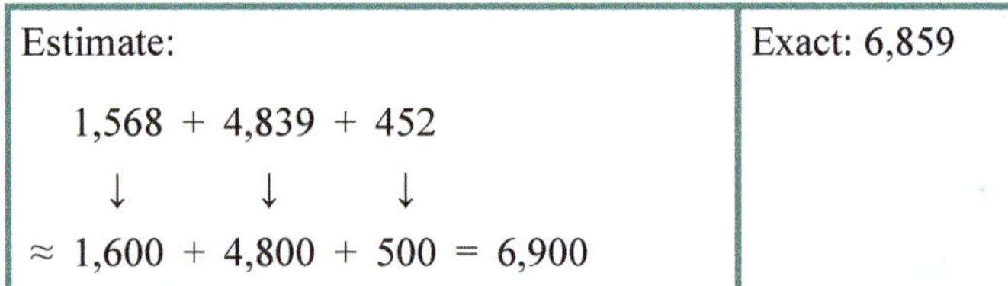

4. a. 3,998; 3,960; 3,991
   b. 6,990; 9,970; 991
   c. 1,900; 6,700; 9,400

5. a. 34,268
   b. 800,046
   c. 406,780

6.

a. 3 ft = 36 in 9 ft = 108 in	b. 2 ft 5 in = 29 in 7 ft 8 in = 92 in	c. 9 ft 2 in =110 in 10 ft 11 in = 131 in

7. a. 5,400 = 90 × 60   b. 16 × 20 = 8 × 40
   c. 7 × 49 + 49 = 8 × 49   d. 24,000 = 300 × 80
   e. 7 × 13 = 5 × 13 + 26   f. 1,500 − 500 = 5 × 200

8. a. Estimate: 8 weeks (8 × \$50 = \$400). Exact: 9 weeks, because 9 × \$45 = \$405. He will have \$6 left over.
   b. She needs 230 cm of string, 69 sheets of paper, and 46 egg cartons.

   c. James had 25 marbles.

   ← 100 →
   Greg | James | Mark

Review, p. 71

1.

a.	b.	c.
20 ÷ 10 + 15 = 17 20 × 10 + 15 = 215	(200 + 100) ÷ 5 = 60 200 + 100 ÷ 5 = 220	10 × 12 + 40 ÷ 10 = 124 10 × (12 + 40) ÷ 10 = 52

2.

a. 3,100 ÷ 100 = 31 450 ÷ 10 = 45	b. 240 ÷ 20 = 12 800 ÷ 40 = 20	c. 4,200 ÷ 600 = 7 3,200 ÷ 80 = 40

3.

a.	b.	c.
45 ÷ 6 = 7 R3 46 ÷ 6 = 7 R4	12 ÷ 7 = 1 R5 27 ÷ 8 = 3 R3	31 ÷ 4 = 7 R3 56 ÷ 9 = 6 R2

4. a. 236   b. 188

5. a. 78 R2   b. 474 R1

6. 288 ÷ 4 = 72. Timmy has 72 seashells.

7. a. 70 ÷ 12 = 5 R10. Mark had five full boxes of candles.
   b. One box had ten candles.

8. $38.88 ÷ 4 = $9.72. One yard cost $9.72.

9. (92 + 85 + 89 + 75 + 89) ÷ 5 = 86. John's average score was 86.

10.

Number	13	40	57	135	354	2,380
Divisible by 3			x	x	x	
Divisible by 5		x		x		x
Divisible by 10		x				x

11.

a. Is 7 a factor of 64? No, because it does not divide evenly into 64. OR No, because 64 ÷ 7 = 9 R1; there is a remainder.	b. Is 98 a multiple of 2? Yes, because it is an even number. OR Yes, because 2 × 49 = 98.
c. Is 76 divisible by 8? No, because 76 ÷ 8 = 9 R4; the division is not even.	d. Is 30 a factor of 30? Yes, because 1 × 30 = 30.

12.

a. 87 is composite. 87 = 3 × 29	b. 89 is prime.	c. 91 is composite. 91 = 7 × 13

13.

a. factors: 1, 2, 3, 4, 6, 8, 12, 24	b. factors: 1, 3, 9, 27
c. factors: 1, 2, 3, 6, 11, 22, 33, 66	d. factors: 1, 3, 5, 15, 25, 75

Puzzle corner:
5, 11, 17, 23, 29, 35, 41, 47, 53, 59, 65, 71, 77, 83, 89, 95

Math Mammoth has a variety of resources to fit your needs. All are available as economical downloads, and most also as printed copies.

- **Math Mammoth Light Blue Series**
  A complete curriculum for grades 1-7. Each grade level includes two student worktexts (A and B), which contain all the instruction and exercises all in the same book, answer keys, tests, cumulative reviews, and a worksheet maker. International (all metric), Canadian, and South African versions are also available.
  https://www.MathMammoth.com/complete-curriculum
  https://www.MathMammoth.com/international/international
  https://www.MathMammoth.com/canada/
  https://www.MathMammoth.com/south_africa/

- **Math Mammoth Skills Review Workbooks**
  These workbooks are intended to be used alongside the Light Blue series full curriculum, and they provide additional review to the topics studied in the main curriculum, in a spiral manner.
  https://www.MathMammoth.com/skills_review_workbooks/

- **Math Mammoth Blue Series**
  Blue Series books are topical worktexts for grades 1-7, containing both instruction and exercises. The topics cover all elementary mathematics from 1st through 7th grade. These books are not tied to grade levels, and are thus great for filling in gaps.
  https://www.MathMammoth.com/blue-series

- **Make It Real Learning**
  These activity workbooks concentrate on answering the question, "Where is math used in real life?" The series includes various workbooks for grades 3-12.
  https://www.MathMammoth.com/worksheets/mirl/

- **Review Workbooks**
  Workbooks for grades 1-7 that provide a comprehensive review of one grade level of math—for example, for review during school break or summer vacation.
  https://www.MathMammoth.com/review_workbooks/

Free gift!

- Receive over 350 free sample pages and worksheets from my books, plus other freebies:
  https://www.MathMammoth.com/worksheets/free

Lastly...

- Inspire4 is an inspirational website for the whole family I've been privileged to help with:
  https://www.inspire4.com

www.ingramcontent.com/pod-product-compliance
Lightning Source LLC
LaVergne TN
LVHW080327110826
845155LV00026B/211

* 9 7 8 1 9 5 4 3 5 8 7 2 0 *